MATH BIBLE QUIZZES

GENESIS TO REVELATION IN MATH

Written: OCTOBER 2019
Authored by: CHRISTINE MARCIA SMALL
Copyright: CHRISTINE MARCIA SMALL ©

TABLE OF CONTENTS

INTRODUCTION

This is a Math Bible quizzes book that gives you the opportunity to search the Bible from Genesis to Revelation Mathematically while learning Bible verses. This can be used for Bible Drill competitions, Bible quizzes, family fun activity, Youth Fellowship and AY programs, Icebreakers for Bible class and many others. There are additions, subtractions, multiplications and divisions questions to be answered. An answer sheet is also provided.

Examples
Book: $6 + 5 = 11$
Chapter: $7 + 3 = 10$
Verse: $1 + 0 = 1$
Answer: 1 Kings 10:1

Book: $10 - 5 = 5$
Chapter: $18 - 3 = 15$
Verse: $9 - 2 = 7$
Answer: Deuteronomy 15:7

Book: $3 \times 5 = 15$
Chapter: $2 \times 2 = 4$
Verse: $2 \times 2 = 4$
Answer: Ezra: 4:4

Book: $20 \div 2 = 10$
Chapter: $36 \div 4 = 9$
Verse: $14 \div 2 = 7$
Answer: 2 Samuel 9:7

CHAPTER ONE
OLD TESTAMENT

There are thirty-nine (39) books in the Old Testament.

BOOKS IN NUMERICAL ORDER	BOOKS
1	Genesis
2	Exodus
3	Leviticus
4	Numbers
5	Deuteronomy
6	Joshua
7	Judges
8	Ruth
9	1 Samuel
10	2 Samuel
11	1 Kings
12	2 Kings
13	1 Chronicles
14	2 Chronicles
15	Ezra
16	Nehemiah
17	Esther
18	Job
19	Psalms
20	Proverbs
21	Ecclesiastes
22	Song of Solomon
23	Isaiah
24	Jeremiah
25	Lamentations
26	Ezekiel
27	Daniel
28	Hosea
29	Joel
30	Amos
31	Obadiah
32	Jonah
33	Micah
34	Nahum
35	Habakkuk
36	Zephaniah
37	Haggai
38	Zechariah
39	Malachi

ADDITIONS

Old Testament

1. Book: $4 + 2 =$
   Chapter: $0 + 1 =$
   Verse: $7 + 2 =$
   Answer: _____

2. Book: $1 + 0 =$
   Chapter: $1 + 0 =$
   Verse: $20 + 7 =$
   Answer: _____

3. Book: $6 + 6 =$
   Chapter: $3 + 2 =$
   Verse: $9 + 5 =$
   Answer: _____

4. Book: $2 + 0 =$
   Chapter: $18 + 2 =$
   Verse: $6 + 9 =$
   Answer: _____

5. Book: $5 + 4 =$
   Chapter: $7 + 8 =$
   Verse: $11 + 11 =$
   Answer: _____

6. Book: $5 + 2 =$
   Chapter: $6 + 1 =$
   Verse: $2 + 2 =$
   Answer: _____

7. Book: $10 + 7 =$
   Chapter: $4 + 0 =$
   Verse: $10 + 6 =$
   Answer: _____

8. Book: $1 + 2 =$
   Chapter: $9 + 2 =$
   Verse: $4 + 3 =$
   Answer: _____

9. Book: $8 + 8 =$
   Chapter: $8 + 1 =$
   Verse: $19 + 3 =$
   Answer: _____

10. Book: $3 + 1 =$
    Chapter: $3 + 30 =$
    Verse: $21 + 33 =$
    Answer: _____

11. Book: $13 + 10 =$
    Chapter: $40 + 25 =$
    Verse: $13 + 11 =$
    Answer: _____

12. Book: $4 + 1 =$
    Chapter: $3 + 2 =$
    Verse: $10 + 7 =$
    Answer: _____

13. Book: $13 + 6 =$
    Chapter: $22 + 5 =$
    Verse: $1 + 6 =$
    Answer: _____

14. Book: $4 + 4 =$

Chapter: $0 + 1 =$

Verse: $20 + 2 =$

Answer: _____

15. Book: $20 + 11 =$

Chapter: $1 + 0 =$

Verse: $15 + 2 =$

Answer: _____

16. Book: $7 + 3 =$

Chapter: $8 + 6 =$

Verse: $12 + 7 =$

Answer: _____

17. Book: $22 + 16 =$

Chapter: $4 + 0 =$

Verse: $3 + 3 =$

Answer: _____

18. Book: $10 + 8 =$

Chapter: $12 + 1 =$

Verse: $4 + 11 =$

Answer: _____

19. Book: $30 + 4 =$

Chapter: $0 + 1 =$

Verse: $2 + 1 =$

Answer: _____

20. Book: $6 + 5 =$

Chapter: $7 + 3 =$

Verse: $1 + 0 =$

Answer: _____

21. Book: $13 + 14 =$

Chapter: $0 + 1 =$

Verse: $16 + 4 =$

Answer: _____

22. Book: $9 + 4 =$

Chapter: $15 + 2 =$

Verse: $1 + 1 =$

Answer: _____

23. Book: $22 + 17 =$

Chapter: $3 + 0 =$

Verse: $0 + 1 =$

Answer: _____

24. Book: $13 + 13 =$

Chapter: $29 + 7 =$

Verse: $13 + 14 =$

Answer: _____

25. Book: $17 + 3 =$

Chapter: $13 + 9 =$

Verse: $1 + 0 =$

Answer: _____

26. Book: $9 + 5 =$

Chapter: $8 + 2 =$

Verse: $11 + 3 =$

Answer: _____

27. Book: 20 + 8 =
    Chapter: 10 + 2 =
    Verse: 5 + 1 =
    Answer: _____

28. Book: 30 + 2 =
    Chapter: 1 + 3 =
    Verse: 10 + 1 =
    Answer: _____

29. Book: 29 + 8 =
    Chapter: 0 + 1 =
    Verse: 5 + 0 =
    Answer: _____

30. Book: 25 + 5 =
    Chapter: 2 + 1 =
    Verse: 0 + 3 =
    Answer: _____

31. Book: 28 + 7 =
    Chapter: 2 + 0 =
    Verse: 8 + 7 =
    Answer: _____

32. Book: 20 + 13 =
    Chapter: 3 + 4 =
    Verse: 5 + 2 =
    Answer: _____

33. Book: 2 + 12 =
    Chapter: 9 + 8 =
    Verse: 18 + 2 =

Answer: _____

34. Book: 10 + 11 =
    Chapter: 1 + 6 =
    Verse: 5+ 4 =
    Answer: _____

35. Book: 14 + 1 =
    Chapter: 2 + 4 =
    Verse: 7 + 5 =
    Answer: _____

36. Book: 16 + 6=
    Chapter: 1 + 2 =
    Verse: 10 + 1 =
    Answer: _____

37. Book: 30 + 6 =
    Chapter: 1 + 1 =
    Verse: 14 + 1 =
    Answer: _____

38. Book: 20 + 9 =
    Chapter: 0 + 1 =
    Verse: 7 + 7 =
    Answer: _____

39. Book: 20 + 5 =
    Chapter: 4 + 1 =
    Verse: 14 + 5 =
    Answer: _____

40. Book: $18 + 1 =$
    Chapter: $7 + 2 =$
    Verse: $1 + 0 =$
    Answer: _____

41. Book: $19 + 4 =$
    Chapter: $20 + 38 =$
    Verse: $4 + 2 =$
    Answer: _____

42. Book: $3 + 2 =$
    Chapter: $7 + 0 =$
    Verse: $11 + 1 =$
    Answer: _____

43. Book: $16 + 4 =$
    Chapter: $20 + 9 =$
    Verse: $10 + 4 =$
    Answer: _____

44. Book: $9 + 9 =$
    Chapter: $14 + 14 =$
    Verse: $19 + 9 =$
    Answer: _____

45. Book: $12 + 14 =$
    Chapter: $22 + 22 =$
    Verse: $15 + 8 =$
    Answer: _____

46. Book: $12 + 12 =$
    Chapter: $21 + 4 =$
    Verse: $10 + 7 =$
    Answer: _____

47. Book: $1 + 1 =$
    Chapter: $30 + 2 =$
    Verse: $6 + 6 =$
    Answer: _____

48. Book: $20 + 4 =$
    Chapter: $25 + 25 =$
    Verse: $24 + 10 =$
    Answer: _____

49. Book: $2 + 2 =$
    Chapter: $20 + 2 =$
    Verse: $14 + 14 =$
    Answer: _____

50. Book: $19 + 6 =$
    Chapter: $0 + 1 =$
    Verse: $19 + 1 =$
    Answer: _____

SUBTRACTIONS

Old Testament

1. Book: $39 - 20 =$
    Chapter: $40 - 6 =$
    Verse: $2 - 1 =$
    Answer: _____

2. Book: $6 - 5 =$
    Chapter: $8 - 6 =$
    Verse: $3 - 1 =$
    Answer: _____

3. Book: $30 - 7 =$
    Chapter: $50 - 13 =$
    Verse: $32 - 12 =$
    Answer: _____

4. Book: $11 - 8 =$
    Chapter: $29 - 10 =$
    Verse: $50 - 32 =$
    Answer: _____

5. Book: $10 - 7 =$
    Chapter: $50 - 30 =$
    Verse: $42 - 14 =$
    Answer: _____

6. Book: $18 - 13 =$
    Chapter: $20 - 6 =$
    Verse: $12 - 4 =$
    Answer: _____

7. Book: $20 - 2 =$
    Chapter: $50 - 8 =$
    Verse: $15 - 3 =$
    Answer: _____

8. Book: $26 - 2 =$
    Chapter: $15 - 8 =$
    Verse: $21 - 10 =$
    Answer: _____

9. Book: $25 - 12 =$
    Chapter: $16 - 5 =$
    Verse: $2 - 1 =$
    Answer: _____

10. Book: $29 - 3 =$
    Chapter: $30 - 10 =$
    Verse: $43 - 16 =$
    Answer: _____

11. Book: $50 - 11 =$
    Chapter: $6 - 3 =$
    Verse: $14 - 8 =$
    Answer: _____

12. Book: $26 - 5 =$
    Chapter: $23 - 11 =$
    Verse: $36 - 23 =$
    Answer: _____

13. Book: $13 - 4 =$
    Chapter: $31 - 13 =$
    Verse: $25 - 11 =$
    Answer: _____

14. Book: $55 - 35 =$
    Chapter: $35 - 15 =$
    Verse: $23 - 1 =$
    Answer: _____

15. Book: $39 - 24 =$
    Chapter: $12 - 9 =$
    Verse: $18 - 7 =$
    Answer: _____

16. Book: $39 - 25 =$
    Chapter: $20 - 6 =$
    Verse: $50 - 21 =$
    Answer: _____

17. Book: $39 - 12 =$
    Chapter: $10 - 4 =$
    Verse: $28 - 1 =$
    Answer: _____

18. Book: $39 - 35 =$
    Chapter: $14 - 3 =$
    Verse: $23 - 8 =$
    Answer: _____

19. Book: $39 - 1 =$
    Chapter: $14 - 4 =$
    Verse: $20 - 8 =$
    Answer: _____

20. Book: $26 - 19 =$
    Chapter: $40 - 24 =$
    Verse: $32 - 14 =$
    Answer: _____

21. Book: $36 - 11 =$
    Chapter: $42 - 39 =$
    Verse: $75 - 11 =$
    Answer: _____

22. Book: $39 - 7 =$
    Chapter: $2 - 1 =$
    Verse: $10 - 7 =$
    Answer: _____

23. Book: $39 - 33 =$
    Chapter: $15 - 5 =$
    Verse: $18 - 6 =$
    Answer: _____

24. Book: $39 - 29 =$
    Chapter: $26 - 4 =$
    Verse: $50 - 3 =$
    Answer: _____

25. Book: $20 - 4 =$
    Chapter: $14 - 13 =$
    Verse: $22 - 11 =$
    Answer: _____

26. Book: $39 - 27 =$
    Chapter: $27 - 4 =$
    Verse: $8 - 5 =$
    Answer: _____

    Answer: _____

27. Book: $39 - 9 =$
    Chapter: $18 - 13 =$
    Verse: $30 - 6 =$
    Answer:
    _____

28. Book: $39 - 22 =$
    Chapter: $19 - 16 =$
    Verse: $7 - 1 =$
    Answer: _____

29. Book: $39 - 6 =$
    Chapter: $18 - 16 =$
    Verse: $4 - 3 =$
    Answer: _____

30. Book: $20 - 10 =$
    Chapter: $23 - 11 =$
    Verse: $5 - 1 =$

MULTIPLICATIONS

Old Testament

1. Book: $2 \times 19 =$
   Chapter: $2 \times 1 =$
   Verse: $4 \times 2 =$
   Answer _____

2. Book: $23 \times 1 =$
   Chapter: $5 \times 1 =$
   Verse: $3 \times 7 =$
   Answer: _____

3. Book: $7 \times 2 =$
   Chapter: $1 \times 1 =$
   Verse: $5 \times 2 =$
   Answer: _____

4. Book: $19 \times 1 =$
   Chapter: $49 \times 2 =$
   Verse: $2 \times 2 =$
   Answer: _____

5. Book: $1 \times 4 =$
   Chapter: $6 \times 3 =$
   Verse: $3 \times 7 =$
   Answer: _____

6. Book: $23 \times 1 =$
   Chapter: $3 \times 3 =$
   Verse: $2 \times 3 =$
   Answer: _____

7. Book: $3 \times 1 =$
   Chapter: $6 \times 4 =$
   Verse: $2 \times 1 =$
   Answer: _____

8. Book: $8 \times 4 =$
   Chapter: $2 \times 1 =$
   Verse: $5 \times 2 =$
   Answer: _____

9. Book: $3 \times 2 =$
   Chapter: $2 \times 1 =$
   Verse: $3 \times 3 =$
   Answer: _____

10. Book: $1 \times 1 =$
    Chapter: $1 \times 1 =$
    Verse: $29 \times 1 =$
    Answer: _____

11. Book: $2 \times 12 =$
    Chapter: $5 \times 2 =$
    Verse: $4 \times 3 =$
    Answer: _____

12. Book: $3 \times 7 =$
    Chapter: $2 \times 2 =$
    Verse: $3 \times 2 =$
    Answer: _____

13. Book: $4 \times 3 =$
Chapter: $1 \times 2 =$
Verse: $11 \times 1 =$
Answer: _____

14. Book: $1 \times 19 =$
Chapter: $5 \times 5 =$
Verse: $4 \times 5 =$
Answer: _____

15. Book: $27 \times 1 =$
Chapter: $1 \times 2 =$
Verse: $23 \times 1 =$
Answer: _____

16. Book: $5 \times 2 =$
Chapter: $2 \times 2 =$
Verse: $2 \times 2 =$
Answer: _____

17. Book: $5 \times 7 =$
Chapter: $2 \times 1 =$
Verse: $2 \times 2 =$
Answer: _____

18. Book: $1 \times 2 =$
Chapter: $2 \times 1 =$
Verse: $5 \times 2 =$
Answer: _____

19. Book: $2 \times 2 =$
Chapter: $13 \times 2 =$
Verse: $3 \times 18 =$

Answer: _____

20. Book: $7 \times 4 =$
Chapter: $2 \times 3 =$
Verse: $3 \times 1 =$
Answer: _____

21. Book: $2 \times 17 =$
Chapter: $2 \times 1 =$
Verse: $4 \times 2 =$
Answer: _____

22. Book: $2 \times 10 =$
Chapter: $1 \times 3 =$
Verse: $13 \times 1 =$
Answer: _____

23. Book: $5 \times 1 =$
Chapter: $2 \times 17 =$
Verse: $1 \times 7 =$
Answer: _____

24. Book: $2 \times 13 =$
Chapter: $43 \times 1 =$
Verse: $1 \times 5 =$
Answer: _____

25. Book: $1 \times 7 =$
Chapter: $5 \times 1 =$
Verse: $2 \times 1 =$
Answer: _____

26. Book: $1 \times 29 =$
    Chapter: $2 \times 1 =$
    Verse: $1 \times 13 =$
    Answer: _____

27. Book: $6 \times 3 =$
    Chapter: $2 \times 19 =$
    Verse: $2 \times 2 =$
    Answer: _____

28. Book: $11 \times 1 =$
    Chapter: $2 \times 11 =$
    Verse: $1 \times 23 =$
    Answer: _____

29. Book: $3 \times 3 =$
    Chapter: $6 \times 5 =$
    Verse: $5 \times 2 =$
    Answer: _____

30. Book: $4 \times 2 =$
    Chapter: $1 \times 2 =$
    Verse: $3 \times 6 =$
    Answer: _____

DIVISIONS
Old Testament

1. Book: $230 \div 10 =$
   Chapter: $52 \div 2 =$
   Verse: $9 \div 3 =$
   Answer _____

2. Book: $380 \div 10 =$
   Chapter: $36 \div 3 =$
   Verse: $24 \div 3 =$
   Answer: _____

3. Book: $42 \div 3 =$
   Chapter: $30 \div 5 =$
   Verse: $32 \div 2 =$
   Answer: _____

4. Book: $100 \div 25 =$
   Chapter: $70 \div 5 =$
   Verse: $96 \div 12 =$
   Answer: _____

5. Book: $96 \div 4 =$
   Chapter: $87 \div 3 =$
   Verse: $33 \div 3 =$
   Answer: _____

6. Book: $42 \div 14 =$
   Chapter: $36 \div 2 =$
   Verse: $9 \div 3 =$
   Answer: _____

7. Book: $96 \div 3 =$
   Chapter: $8 \div 8 =$
   Verse: $34 \div 2 =$
   Answer: _____

8. Book: $42 \div 7 =$
   Chapter: $27 \div 9 =$
   Verse: $28 \div 7 =$
   Answer: _____

9. Book: $42 \div 2 =$
   Chapter: $15 \div 3 =$
   Verse: $4 \div 4 =$
   Answer: _____

10. Book: $3 \div 3 =$
    Chapter: $21 \div 7 =$
    Verse: $12 \div 2 =$
    Answer: _____

11. Book: $24 \div 2 =$
    Chapter: $25 \div 5 =$
    Verse: $52 \div 2 =$
    Answer: _____

12. Book: $80 \div 4 =$
    Chapter: $96 \div 6 =$
    Verse: $180 \div 20 =$
    Answer: _____

13. Book: $54 \div 2 =$
    Chapter: $45 \div 9 =$
    Verse: $51 \div 3 =$

Answer: _____

14. Book: $40 \div 4 =$
Chapter: $30 \div 5 =$
Verse: $36 \div 3 =$
Answer: _____

15. Book: $350 \div 10 =$
Chapter: $45 \div 15 =$
Verse: $60 \div 6 =$
Answer: _____

16. Book: $30 \div 15 =$
Chapter: $16 \div 4 =$
Verse: $15 \div 5 =$
Answer: _____

17. Book: $32 \div 4 =$
Chapter: $1 \div 1 =$
Verse: $160 \div 10 =$
Answer: _____

18. Book: $56 \div 2 =$
Chapter: $42 \div 3 =$
Verse: $7 \div 7 =$
Answer: _____

19. Book: $102 \div 3 =$
Chapter: $30 \div 10 =$
Verse: $42 \div 6 =$
Answer: _____

20. Book: $60 \div 12 =$
Chapter: $124 \div 4$
Verse: $84 \div 7 =$
Answer: _____

21. Book: $52 \div 2 =$
Chapter: $39 \div 3 =$
Verse: $21 \div 7 =$
Answer: _____

22. Book: $49 \div 7 =$
Chapter: $44 \div 4 =$
Verse: $30 \div 6 =$
Answer: _____

23. Book: $87 \div 3 =$
Chapter: $18 \div 9 =$
Verse: $46 \div 2 =$
Answer: _____

24. Book: $36 \div 2 =$
Chapter: $10 \div 5 =$
Verse: $6 \div 6 =$
Answer: _____

25. Book: $44 \div 4 =$
Chapter: $66 \div 3 =$
Verse: $86 \div 2 =$
Answer: _____

26. Book: $45 \div 5 =$
Chapter: $81 \div 3 =$
Verse: $11 \div 11 =$

Answer: _____

27. Book: $48 \div 3 =$
Chapter: $20 \div 5 =$
Verse: $9 \div 9 =$
Answer: _____

28. Book: $120 \div 4 =$
Chapter: $40 \div 5 =$
Verse: $77 \div 7 =$
Answer: _____

29. Book: $80 \div 5 =$
Chapter: $54 \div 9 =$
Verse: $160 \div 10 =$
Answer: _____

30. Book: $36 \div 3 =$
Chapter: $40 \div 10 =$
Verse: $35 \div 5 =$
Answer: _____

CHAPTER 2
NEW TESTAMENT

There are twenty-seven books in the New Testament.

BOOKS POSITIONS	BOOKS
1	Matthew
2	Mark
3	Luke
4	John
5	Acts
6	Romans
7	1 Corinthians
8	2 Corinthians
9	Galatians
10	Ephesians
11	Philippians
12	Colossians
13	1 Thessalonians
14	2 Thessalonians
15	1 Timothy
16	2 Timothy
17	Titus
18	Philemon
19	Hebrews
20	James
21	1 Peter
22	2 Peter
23	1 John
24	2 John
25	3 John
26	Jude
27	Revelation

ADDITIONS
New Testament

1. Book: $2 + 0 =$
    Chapter: $9 + 4 =$
    Verse: $17 + 18 =$
    Answer: _____

2. Book: $3 + 1 =$
    Chapter: $11 + 3 =$
    Verse: $2 + 1 =$
    Answer: _____

3. Book: $2 + 4 =$
    Chapter: $2 + 3 =$
    Verse: $1 + 4 =$
    Answer: _____

4. Book: $3 + 5 =$
    Chapter: $1 + 4 =$
    Verse: $8 + 2 =$
    Answer: _____

5. Book: $1 + 0 =$
    Chapter: $3 + 4 =$
    Verse: $0 + 1 =$
    Answer: _____

6. Book: $7 + 2 =$
    Chapter: $2 + 3 =$
    Verse: $8 + 7 =$
    Answer: _____

7. Book: $12 + 1 =$
    Chapter: $5 + 0 =$
    Verse: $10 + 12 =$
    Answer: _____

8. Book: $8 + 5 =$
    Chapter: $1 + 0 =$
    Verse: $7 + 2 =$
    Answer: _____

9. Book: $3 + 0 =$
    Chapter: $0 + 1 =$
    Verse: $20 + 22 =$
    Answer: _____

10. Book: $3 + 2 =$
    Chapter: $1 + 4 =$
    Verse: $17 + 2 =$
    Answer: _____

11. Book: $6 + 5 =$
    Chapter: $1 + 0 =$
    Verse: $7 + 4 =$
    Answer: _____

12. Book: $1 + 13 =$
    Chapter: $0 + 1 =$
    Verse: $2 + 2 =$
    Answer: _____

13. Book: $2 + 14 =$
    Chapter: $1 + 1 =$
    Verse: $14 + 2 =$

Answer: _____

14. Book: $9 + 1 =$
    Chapter: $1 + 1 =$
    Verse: $5 + 3 =$
    Answer: _____

15. Book: $8 + 9 =$
    Chapter: $0 + 1 =$
    Verse: $6 + 9 =$
    Answer: _____

16. Book: $4 + 15 =$
    Chapter: $6 + 7 =$
    Verse: $1 + 1 =$
    Answer: _____

17. Book: $17 + 4 =$
    Chapter: $0 + 1 =$
    Verse: $16 + 6 =$
    Answer: _____

18. Book: $2 + 0 =$
    Chapter: $3 + 3 =$
    Verse: $27 + 6 =$
    Answer: _____

19. Book: $15 + 3 =$
    Chapter: $0 + 1 =$
    Verse: $7 + 11 =$
    Answer: _____

20. Book: $10 + 12 =$
    Chapter: $0 + 1 =$
    Verse: $16 + 4 =$
    Answer: _____

21. Book: $16 + 7 =$
    Chapter: $0 + 1 =$
    Verse: $5 + 5 =$
    Answer: _____

22. Book: $15 + 12 =$
    Chapter: $2 + 1 =$
    Verse: $11 + 9 =$
    Answer: _____

23. Book: $0 + 1 =$
    Chapter: $3 + 2 =$
    Verse: $6 + 1 =$
    Answer: _____

24. Book: $6 + 6 =$
    Chapter: $1 + 0 =$
    Verse: $6 + 3 =$
    Answer: _____

25. Book: $16 + 4 =$
    Chapter: $1 + 0 =$
    Verse: $17 + 5 =$
    Answer: _____

26. Book: $18 + 6 =$
    Chapter: $0 + 1 =$
    Verse: $3 + 3 =$

Answer: _____

27. Book: $8 + 6 =$
    Chapter: $2 + 1 =$
    Verse: $9 + 7 =$
    Answer: _____

28. Book: $18 + 1 =$
    Chapter: $4 + 2 =$
    Verse: $10 + 2 =$
    Answer: _____

29. Book: $5 + 10 =$
    Chapter: $0 + 1 =$
    Verse: $5 + 3 =$
    Answer: _____

30. Book: $6 + 1 =$
    Chapter: $1 + 1 =$
    Verse: $9 + 3 =$
    Answer: _____

SUBTRACTIONS

New Testament

1. Book: $9 - 5 =$
   Chapter: $6 - 2 =$
   Verse: $50 - 8 =$
   Answer _____

2. Book: $25 - 19 =$
   Chapter: $10 - 3 =$
   Verse: $19 - 7 =$
   Answer: _____

3. Book: $15 - 6 =$
   Chapter: $8 - 2 =$
   Verse: $22 - 15 =$
   Answer: _____

4. Book: $6 - 5 =$
   Chapter: $27 - 13 =$
   Verse: $36 - 22 =$
   Answer: _____

5. Book: $17 - 12 =$
   Chapter: $20 - 15 =$
   Verse: $30 - 1 =$
   Answer: _____

6. Book: $50 - 42 =$
   Chapter: $20 - 12 =$
   Verse: $24 - 10 =$
   Answer: _____

7. Book: $16 - 5 =$
   Chapter: $6 - 4 =$
   Verse: $5 - 3 =$
   Answer: _____

8. Book: $16 - 13 =$
   Chapter: $17 - 11 =$
   Verse: $36 - 9 =$
   Answer: _____

9. Book: $20 - 8 =$
   Chapter: $5 - 2 =$
   Verse: $7 - 5 =$
   Answer: _____

10. Book: $20 - 4 =$
    Chapter: $16 - 15 =$
    Verse: $18 - 11 =$
    Answer: _____

11. Book: $26 - 16 =$
    Chapter: $10 - 6 =$
    Verse: $32 - 3 =$
    Answer: _____

12. Book: $20 - 3 =$
    Chapter: $6 - 4 =$
    Verse: $8 - 2 =$
    Answer: _____

13. Book: $3 - 2 =$
    Chapter: $20 - 13 =$
    Verse: $29 - 11 =$
    Answer: _____

14. Book: $17 - 4 =$
    Chapter: $3 - 1 =$
    Verse: $20 - 7 =$
    Answer: _____

15. Book: $24 - 9 =$
    Chapter: $2 - 1 =$
    Verse: $36 - 24 =$
    Answer: _____

16. Book: $27 - 8 =$
    Chapter: $33 - 21 =$
    Verse: $18 - 17 =$
    Answer: _____

17. Book: $27 - 6 =$
    Chapter: $12 - 10 =$
    Verse: $16 - 7 =$
    Answer: _____

18. Book: $35 - 12 =$
    Chapter: $4 - 2 =$
    Verse: $10 - 6 =$
    Answer: _____

19. Book: $16 - 14 =$
    Chapter: $20 - 13 =$
    Verse: $16 - 7 =$

    Answer: _____

20. Book: $18 - 4 =$
    Chapter: $7 - 6 =$
    Verse: $16 - 5 =$
    Answer: _____

21. Book: $27 - 7 =$
    Chapter: $2 - 1 =$
    Verse: $40 - 23 =$
    Answer: _____

22. Book: $64 - 42 =$
    Chapter: $7 - 5 =$
    Verse: $40 - 18 =$
    Answer: _____

23. Book: $66 - 39 =$
    Chapter: $15 - 12 =$
    Verse: $31 - 10 =$
    Answer: _____

24. Book: $45 - 21 =$
    Chapter: $2 - 1 =$
    Verse: $8 - 6 =$
    Answer: _____

25. Book: $16 - 10 =$
    Chapter: $20 - 7 =$
    Verse: $19 - 6 =$
    Answer: _____

26. Book: $27 - 9 =$
    Chapter: $5 - 4 =$
    Verse: $30 - 5 =$
    Answer: _____

27. Book: $13 - 3 =$
    Chapter: $10 - 6 =$
    Verse: $30 - 4 =$
    Answer: _____

28. Book: $20 - 3 =$
    Chapter: $8 - 6 =$
    Verse: $15 - 2 =$
    Answer: _____

29. Book: $3 - 1 =$
    Chapter: $14 - 3 =$
    Verse: $30 - 6 =$
    Answer: _____

30. Book: $25 - 18 =$
    Chapter: $16 - 3 =$
    Verse: $1 - 0 =$
    Answer: _____

MULTIPLICATIONS

New Testament

1. Book: $2 \times 1 =$
   Chapter: $3 \times 4 =$
   Verse: $1 \times 1 =$
   Answer _____

2. Book: $5 \times 1 =$
   Chapter: $1 \times 2 =$
   Verse: $5 \times 9 =$
   Answer: _____

3. Book: $7 \times 1 =$
   Chapter: $2 \times 7 =$
   Verse: $5 \times 8 =$
   Answer: _____

4. Book: $1 \times 11 =$
   Chapter: $4 \times 1 =$
   Verse: $13 \times 1 =$
   Answer: _____

5. Book: $3 \times 3 =$
   Chapter: $2 \times 2 =$
   Verse: $5 \times 4 =$
   Answer: _____

6. Book: $13 \times 1 =$
   Chapter: $1 \times 3 =$
   Verse: $3 \times 3 =$
   Answer: _____

7. Book: $5 \times 3 =$
   Chapter: $1 \times 3 =$
   Verse: $2 \times 1 =$
   Answer: _____

8. Book: $4 \times 4 =$
   Chapter: $2 \times 1 =$
   Verse: $4 \times 2 =$
   Answer: _____

9. Book: $3 \times 1 =$
   Chapter: $2 \times 1 =$
   Verse: $5 \times 8 =$
   Answer: _____

10. Book: $3 \times 2 =$
    Chapter: $3 \times 3 =$
    Verse: $4 \times 4 =$
    Answer: _____

11. Book: $1 \times 1 =$
    Chapter: $6 \times 2 =$
    Verse: $5 \times 7 =$
    Answer: _____

12. Book: $2 \times 5 =$
    Chapter: $2 \times 2 =$
    Verse: $5 \times 3 =$
    Answer: _____

13. Book: $3 \times 5 =$
    Chapter: $2 \times 2 =$
    Verse: $3 \times 4 =$
    Answer: _____

14. Book: $2 \times 2 =$
    Chapter: $4 \times 2 =$
    Verse: $7 \times 1 =$
    Answer: _____

15. Book: $2 \times 4 =$
    Chapter: $2 \times 2 =$
    Verse: $1 \times 2 =$
    Answer: _____

16. Book: $11 \times 1 =$
    Chapter: $2 \times 1 =$
    Verse: $7 \times 2 =$
    Answer: _____

17. Book: $2 \times 7 =$
    Chapter: $1 \times 3 =$
    Verse: $5 \times 2 =$
    Answer: _____

18. Book: $17 \times 1 =$
    Chapter: $3 \times 1 =$
    Verse: $2 \times 4 =$
    Answer: _____

19. Book: $11 \times 1 =$
    Chapter: $2 \times 2 =$
    Verse: $4 \times 2 =$
    Answer: _____

20. Book: $19 \times 1 =$
    Chapter: $11 \times 1 =$
    Verse: $1 \times 1 =$
    Answer: _____

21. Book: $3 \times 7 =$
    Chapter: $1 \times 3 =$
    Verse: $4 \times 2 =$
    Answer: _____

22. Book: $13 \times 1 =$
    Chapter: $2 \times 2 =$
    Verse: $4 \times 4 =$
    Answer: _____

23. Book: $2 \times 9 =$
    Chapter: $1 \times 1 =$
    Verse: $2 \times 2 =$
    Answer: _____

24. Book: $2 \times 10 =$
    Chapter: $1 \times 1 =$
    Verse: $4 \times 2 =$
    Answer: _____

25. Book: $3 \times 9 =$
    Chapter: $2 \times 11 =$
    Verse: $4 \times 3 =$
    Answer: _____

26. Book: $3 \times 4 =$
    Chapter: $1 \times 1 =$
    Verse: $3 \times 1 =$
    Answer: _____

27. Book: $2 \times 1 =$
    Chapter: $4 \times 2 =$
    Verse: $4 \times 10 =$
    Answer: _____

28. Book: $2 \times 7 =$
    Chapter: $1 \times 2 =$
    Verse: $13 \times 1 =$
    Answer: _____

29. Book: $10 \times 1 =$
    Chapter: $1 \times 6 =$
    Verse: $2 \times 5 =$
    Answer: _____

30. Book: $1 \times 1 =$
    Chapter: $3 \times 3 =$
    Verse: $6 \times 3 =$
    Answer: _____

DIVISIONS
New Testament

1. Book: $16 \div 4 =$
   Chapter: $48 \div 6 =$
   Verse: $160 \div 5 =$
   Answer _____

2. Book: $42 \div 7 =$
   Chapter: $56 \div 7 =$
   Verse: $42 \div 3 =$
   Answer: _____

3. Book: $26 \div 13 =$
   Chapter: $18 \div 6 =$
   Verse: $80 \div 20 =$
   Answer: _____

4. Book: $14 \div 2 =$
   Chapter: $54 \div 6 =$
   Verse: $26 \div 2 =$
   Answer: _____

5. Book: $45 \div 5 =$
   Chapter: $60 \div 10 =$
   Verse: $9 \div 3 =$
   Answer: _____

6. Book: $144 \div 12 =$
   Chapter: $45 \div 15 =$
   Verse: $75 \div 3 = 25$
   Answer: _____

7. Book: $16 \div 1 =$
   Chapter: $2 \div 1 =$
   Verse: $95 \div 5 =$
   Answer: _____

8. Book: $9 \div 3 =$
   Chapter: $48 \div 8 =$
   Verse: $90 \div 2 =$
   Answer: _____

9. Book: $39 \div 3 =$
   Chapter: $60 \div 12 =$
   Verse: $85 \div 5 =$
   Answer: _____

10. Book: $55 \div 11 =$
    Chapter: $63 \div 7 =$
    Verse: $72 \div 12 =$
    Answer: _____

11. Book: $1 \div 1 =$
    Chapter: $32 \div 2 =$
    Verse: $96 \div 6 =$
    Answer: _____

12. Book: $40 \div 5 =$
    Chapter: $26 \div 2 =$
    Verse: $35 \div 5 =$
    Answer: _____

13. Book: $270 \div 27 =$
    Chapter: $50 \div 10 =$
    Verse: $60 \div 3 =$
    Answer: _____

14. Book: $14 \div 1 =$
    Chapter: $36 \div 12 =$
    Verse: $24 \div 4 =$
    Answer: _____

15. Book: $36 \div 2 =$
    Chapter: $1 \div 1 =$
    Verse: $10 \div 2 =$
    Answer: _____

16. Book: $20 \div 1 =$
    Chapter: $10 \div 10 =$
    Verse: $57 \div 3 =$
    Answer: _____

17. Book: $42 \div 2 =$
    Chapter: $12 \div 3 =$
    Verse: $56 \div 7 =$
    Answer: _____

18. Book: $46 \div 2 =$
    Chapter: $23 \div 23 =$
    Verse: $28 \div 4 =$
    Answer: _____

19. Book: $42 \div 3 =$
    Chapter: $26 \div 13 =$
    Verse: $12 \div 4 =$
    Answer: _____

20. Book: $6 \div 3 =$
    Chapter: $48 \div 3 =$
    Verse: $54 \div 9 =$
    Answer: _____

21. Book: $60 \div 4 =$
    Chapter: $100 \div 50 =$
    Verse: $80 \div 10 =$
    Answer: _____

22. Book: $19 \div 1 =$
    Chapter: $8 \div 2 =$
    Verse: $40 \div 10 =$
    Answer: _____

23. Book: $135 \div 5 =$
    Chapter: $84 \div 6 =$
    Verse: $48 \div 4 =$
    Answer: _____

24. Book: $49 \div 7 =$
    Chapter: $60 \div 15 =$
    Verse: $28 \div 2 =$
    Answer: _____

25. Book: $15 \div 3 =$
    Chapter: $34 \div 2 =$
    Verse: $90 \div 3 =$

Answer: _____

26. Book: $66 \div 6 =$
    Chapter: $4 \div 2 =$
    Verse: $44 \div 4 =$
    Answer: _____

27. Book: $108 \div 4 =$
    Chapter: $147 \div 7 =$
    Verse: $20 \div 5 =$
    Answer: _____

28. Book: $1 \div 1 =$
    Chapter: $100 \div 10 =$
    Verse: $90 \div 3 =$
    Answer: _____

29. Book: $44 \div 2 =$
    Chapter: $30 \div 10 =$
    Verse: $70 \div 5 =$
    Answer: _____

30. Book: $70 \div 7 =$
    Chapter: $44 \div 11 =$
    Verse: $96 \div 3 =$
    Answer: _____

CHAPTER 3
OLD AND NEW TESTAMENT KEY

There are sixty-six books in the Bible.

BOOKS POSITIONS	BOOKS
1	Genesis
2	Exodus
3	Leviticus
4	Numbers
5	Deuteronomy
6	Joshua
7	Judges
8	Ruth
9	1 Samuel
10	2 Samuel
11	1 Kings
12	2 Kings
13	1 Chronicles
14	2 Chronicles
15	Ezra
16	Nehemiah
17	Esther
18	Job
19	Psalms
20	Proverbs
21	Ecclesiastes
22	Song of Solomon
23	Isaiah

24	Jeremiah
25	Lamentations
26	Ezekiel
27	Daniel
28	Hosea
29	Joel
30	Amos
31	Obadiah
32	Jonah
33	Micah
34	Nahum
35	Habakkuk
36	Zephaniah
37	Haggai
38	Zechariah
39	Malachi

40	Matthew
41	Mark
42	Luke
43	John
44	Acts
45	Romans
46	1 Corinthians
47	2 Corinthians
48	Galatians
49	Ephesians
50	Philippians
51	Colossians
52	1 Thessalonians
53	2 Thessalonians
54	1 Timothy
55	2 Timothy
56	Titus

57	Philemon
58	Hebrews
59	James
60	1 Peter
61	2 Peter
62	1 John

63	2 John
64	3 John
65	Jude
66	Revelation

ADDITIONS

Old and New Testament

1. Book: $17 + 6 =$
   Chapter: $5 + 9 =$
   Verse: $3 + 0 =$
   Answer _____

2. Book: $18 + 1 =$
   Chapter: $28 + 43 =$
   Verse: $7 + 1 =$
   Answer: _____

3. Book: $21 + 19 =$
   Chapter: $11 + 7 =$
   Verse: $1 + 2 =$
   Answer: _____

4. Book: $4 + 0 =$
   Chapter: $7 + 5 =$
   Verse: $3 + 3 =$
   Answer: _____

5. Book: $14 + 10 =$
   Chapter: $7 + 2 =$
   Verse: $12 + 11 =$
   Answer: _____

6. Book: $25 + 16 =$
   Chapter: $5 + 9 =$
   Verse: $50 + 22 =$
   Answer: _____

7. Book: $16 + 16 =$
   Chapter: $0 + 3 =$
   Verse: $4 + 1 =$
   Answer: _____

8. Book: $3 + 3 =$
   Chapter: $3 + 2 =$
   Verse: $8 + 7 =$
   Answer: _____

9. Book: $28 + 18 =$
   Chapter: $8 + 4 =$
   Verse: $3 + 9 =$
   Answer: _____

10. Book: $35 + 7 =$
    Chapter: $7 + 1 =$
    Verse: $0 + 1 =$
    Answer: _____

11. Book: $15 + 6 =$
    Chapter: $3 + 4 =$
    Verse: $6 + 3 =$
    Answer: _____

12. Book: $40 + 4 =$
    Chapter: $3 + 0 =$
    Verse: $2 + 4 =$
    Answer: _____

13. Book: $11 + 3 =$
    Chapter: $16 + 13 =$
    Verse: $7 + 8 =$
    Answer: _____

14. Book: $3 + 0 =$
    Chapter: $14 + 9 =$
    Verse: $18 + 4 =$
    Answer: _____

15. Book: $20 + 25 =$
    Chapter: $6 + 6 =$
    Verse: $9 + 9 =$
    Answer: _____

16. Book: $8 + 4 =$
    Chapter: $3 + 5 =$
    Verse: $0 + 1 =$
    Answer: _____

17. Book: $6 + 4 =$
    Chapter: $2 + 7 =$
    Verse: $3 + 4 =$
    Answer: _____

18. Book: $17 + 3 =$
    Chapter: $9 + 5 =$
    Verse: $0 + 1 =$
    Answer: _____

19. Book: $16 + 27 =$
    Chapter: $3 + 2 =$
    Verse: $7 + 1 =$
    Answer: _____

20. Book: $0 + 1 =$
    Chapter: $3 + 1 =$
    Verse: $9 + 2 =$
    Answer: _____

SUBTRACTIONS
Old and New Testament

1. Book: $66 - 26 =$
   Chapter: $27 - 7 =$
   Verse: $40 - 12 =$
   Answer _____

2. Book: $50 - 27 =$
   Chapter: $66 - 25 =$
   Verse: $30 - 13 =$
   Answer: _____

3. Book: $85 - 41 =$
   Chapter: $27 - 21 =$
   Verse: $12 - 9 =$
   Answer: _____

4. Book: $50 - 31 =$
   Chapter: $150 - 64 =$
   Verse: $17 - 6 =$
   Answer: _____

5. Book: $66 - 42 =$
   Chapter: $25 - 12 =$
   Verse: $30 - 7 =$
   Answer: _____

6. Book: $60 - 14 =$
   Chapter: $22 - 9 =$
   Verse: $6 - 5 =$
   Answer: _____

7. Book: $25 - 22 =$
   Chapter: $40 - 14 =$
   Verse: $20 - 7 =$
   Answer: _____

8. Book: $64 - 32 =$
   Chapter: $8 - 4 =$
   Verse: $17 - 15 =$
   Answer: _____

9. Book: $60 - 12 =$
   Chapter: $10 - 8 =$
   Verse: $27 - 7 =$
   Answer: _____

10. Book: $7 - 1 =$
    Chapter: $18 - 12 =$
    Verse: $42 - 22 =$
    Answer: _____

11. Book: $39 - 38 =$
    Chapter: $20 - 3 =$
    Verse: $2 - 1 =$
    Answer: _____

12. Book: $95 - 49 =$
    Chapter: $45 - 30 =$
    Verse: $66 - 14 =$
    Answer: _____

13. Book: $39 - 18 =$
    Chapter: $10 - 2 =$
    Verse: $19 - 8 =$
    Answer: _____

14. Book: $50 - 5 =$
    Chapter: $21 - 8 =$
    Verse: $13 - 3 =$
    Answer: _____

15. Book: $66 - 52 =$
    Chapter: $15 - 14 =$
    Verse: $22 - 10 =$
    Answer: _____

16. Book: $25 - 13 =$
    Chapter: $40 - 28 =$
    Verse: $11 - 2 =$
    Answer: _____

17. Book: $66 - 23 =$
    Chapter: $14 - 8 =$
    Verse: $26 - 12 =$
    Answer: _____

18. Book: $18 - 8 =$
    Chapter: $26 - 10 =$
    Verse: $2 - 1 =$
    Answer: _____

19. Book: $35 - 8 =$
    Chapter: $9 - 3 =$
    Verse: $31 - 9 =$
    Answer: _____

20. Book: $60 - 40 =$
    Chapter: $20 - 12 =$
    Verse: $15 - 4 =$
    Answer: _____

MULTIPLICATIONS

Old and New Testament

1. Book: $41 \times 1 =$
   Chapter: $5 \times 3 =$
   Verse: $23 \times 2 =$
   Answer _____

2. Book: $2 \times 3 =$
   Chapter: $2 \times 2 =$
   Verse: $23 \times 1 =$
   Answer: _____

3. Book: $7 \times 3 =$
   Chapter: $3 \times 3 =$
   Verse: $5 \times 1 =$
   Answer: _____

4. Book: $2 \times 21 =$
   Chapter: $2 \times 2 =$
   Verse: $4 \times 2 =$
   Answer: _____

5. Book: $7 \times 2 =$
   Chapter: $3 \times 8 =$
   Verse: $4 \times 5 =$
   Answer: _____

6. Book: $2 \times 2 =$
   Chapter: $8 \times 4 =$
   Verse: $9 \times 3 =$
   Answer: _____

7. Book: $23 \times 1 =$
   Chapter: $3 \times 4 =$
   Verse: $1 \times 2 =$
   Answer: _____

8. Book: $4 \times 10 =$
   Chapter: $5 \times 5 =$
   Verse: $2 \times 5 =$
   Answer: _____

9. Book: $2 \times 19 =$
   Chapter: $4 \times 1 =$
   Verse: $2 \times 3 =$
   Answer: _____

10. Book: $6 \times 4 =$
    Chapter: $2 \times 1 =$
    Verse: $1 \times 7 =$
    Answer: _____

11. Book: $2 \times 16 =$
    Chapter: $3 \times 1 =$
    Verse: $1 \times 3 =$
    Answer: _____

12. Book: $2 \times 22 =$
    Chapter: $4 \times 1 =$
    Verse: $6 \times 2 =$
    Answer: _____

13. Book: $6 \times 2 =$
    Chapter: $1 \times 19 =$
    Verse: $5 \times 3 =$
    Answer: _____

14. Book: $19 \times 1 =$
    Chapter: $8 \times 15 =$
    Verse: $1 \times 1 =$
    Answer: _____

15. Book: $13 \times 5 =$
    Chapter: $1 \times 1 =$
    Verse: $7 \times 3 =$
    Answer: _____

16. Book: $2 \times 23 =$
    Chapter: $4 \times 2 =$
    Verse: $3 \times 2 =$
    Answer: _____

17. Book: $5 \times 2 =$
    Chapter: $3 \times 7 =$
    Verse: $7 \times 1 =$
    Answer: _____

18. Book: $8 \times 8 =$
    Chapter: $1 \times 1 =$
    Verse: $2 \times 1 =$
    Answer: _____

19. Book: $43 \times 1 =$
    Chapter: $1 \times 11 =$
    Verse: $22 \times 2 =$
    Answer: _____

20. Book: $1 \times 1 =$
    Chapter: $8 \times 3 =$
    Verse: $17 \times 3 =$
    Answer: _____

DIVISIONS

Old and New Testament

1. Book: $230 \div 10 =$
   Chapter: $52 \div 2 =$
   Verse: $16 \div 4 =$
   Answer _____

2. Book: $82 \div 2 =$
   Chapter: $30 \div 3 =$
   Verse: $57 \div 3 =$
   Answer: _____

3. Book: $48 \div 2 =$
   Chapter: $64 \div 4 =$
   Verse: $17 \div 1 =$
   Answer: _____

4. Book: $36 \div 6 =$
   Chapter: $40 \div 5 =$
   Verse: $36 \div 2 =$
   Answer: _____

5. Book: $80 \div 2 =$
   Chapter: $56 \div 2 =$
   Verse: $36 \div 6 =$
   Answer: _____

6. Book: $56 \div 4 =$
   Chapter: $21 \div 3 =$
   Verse: $28 \div 2 =$
   Answer: _____

7. Book: $64 \div 2 =$
   Chapter: $21 \div 7 =$
   Verse: $30 \div 3$
   Answer: _____

8. Book: $43 \div 1 =$
   Chapter: $60 \div 4 =$
   Verse: $70 \div 10 =$
   Answer: _____

9. Book: $48 \div 8 =$
   Chapter: $72 \div 3 =$
   Verse: $62 \div 2 =$
   Answer: _____

10. Book: $1 \div 1 =$
    Chapter: $82 \div 2 =$
    Verse: $100 \div 4 =$
    Answer: _____

11. Book: $92 \div 2 =$
    Chapter: $270 \div 27 =$
    Verse: $52 \div 4 =$
    Answer: _____

12. Book: $76 \div 4 =$
    Chapter: $236 \div 2 =$
    Verse: $32 \div 4 =$
    Answer: _____

13. Book: $63 \div 3 =$
    Chapter: $20 \div 2 =$
    Verse: $33 \div 3 =$
    Answer: _____

14. Book: $144 \div 12 =$
    Chapter: $100 \div 5 =$
    Verse: $40 \div 8 =$
    Answer: _____

15. Book: $132 \div 3 =$
    Chapter: $64 \div 8 =$
    Verse: $85 \div 5 =$
    Answer: _____

16. Book: $36 \div 9 =$
    Chapter: $108 \div 3 =$
    Verse: $54 \div 6 =$
    Answer: _____

17. Book: $108 \div 4 =$
    Chapter: $21 \div 3 =$
    Verse: $150 \div 6 =$
    Answer: _____

18. Book: $84 \div 2 =$
    Chapter: $45 \div 9 =$
    Verse: $40 \div 2 =$
    Answer: _____

19. Book: $60 \div 3 =$
    Chapter: $62 \div 2 =$
    Verse: $30 \div 3 =$
    Answer: _____

20. Book: $95 \div 5 =$
    Chapter: $438 \div 3 =$
    Verse: $40 \div 5 =$
    Answer: _____

CHAPTER FOUR

Old Testament Additions

1. Book: 4 + 2 = 6
   Chapter: 0 + 1 = 1
   Verse: 7 + 2 = 9
   Answer   Joshua 6:9

2. Book: 1 + 0 = 1
   Chapter: 1 + 0 = 1
   Verse: 20 + 7= 27
   Answer: Genesis 1:27

3. Book: 6 + 6 = 12
   Chapter: 3 + 2 = 5
   Verse: 9 + 5 = 14
   Answer: 2 Kings 5:14

4. Book: 2 + 0 = 2
   Chapter: 18 + 2 = 20
   Verse: 6 + 9 = 15
   Answer: Exodus 20:15

5. Book: 5 + 4 = 9
   Chapter: 7 + 8 = 15
   Verse: 11 + 11 = 22
   Answer: 1 Samuel 15:22

6. Book: 5 + 2 = 7
   Chapter: 6 + 1 = 7
   Verse: 2 + 2 = 4
   Answer: Judges 7:4

7. Book: 10 + 7 = 17
   Chapter: 4 + 0 = 4
   Verse: 10 + 6 = 16
   Answer: Esther 4:16

ANSWERS

8. Book: 1 + 2 = 3
   Chapter: 9 + 2 = 11
   Verse: 4 + 3 = 7
   Answer: Leviticus 11:7

9. Book: 8 + 8 = 16
   Chapter: 8 + 1 = 9
   Verse: 19 + 3 = 22
   Answer: Nehemiah 9:22

10. Book: 3 + 1 = 4
    Chapter: 3 + 30 = 33
    Verse: 21 + 33 = 54
    Answer: Numbers 33:54

11. Book: 13 + 10 = 23
    Chapter: 40 + 25 = 65
    Verse: 13 + 11 = 24
    Answer: Isaiah 65:24

12. Book: 4 + 1 = 5
    Chapter: 3 + 2 = 5
    Verse: 10 + 7 = 17
    Answer: Deuteronomy 5:17

13. Book: 13 + 6 = 19
    Chapter: 22 + 5 = 27
    Verse: 1 + 6 = 7
    Answer: Psalm 27:7

14. Book: 4 + 4 = 8
    Chapter: 0 + 1 = 1
    Verse: 20 + 2 = 22
    Answer: Ruth 1:22

15. Book: $20 + 11 = 31$
    Chapter: $1 + 0 = 1$
    Verse: $15 + 2 = 17$
    Answer: Obadiah 1:17

16. Book: $7 + 3 = 10$
    Chapter: $8 + 6 = 14$
    Verse: $12 + 7 = 19$
    Answer: 2 Samuel 14:19

17. Book: $22 + 16 = 38$
    Chapter: $4 + 0 = 4$
    Verse: $3 + 3 = 6$
    Answer: Zechariah 4:6

18. Book: $10 + 8 = 18$
    Chapter: $12 + 1 = 13$
    Verse: $4 + 11 = 15$
    Answer: Job 13:15

19. Book: $30 + 4 = 34$
    Chapter: $0 + 1 = 1$
    Verse: $2 + 1 = 3$
    Answer: Nahum 1:3

20. Book: $6 + 5 = 11$
    Chapter: $7 + 3 = 10$
    Verse: $1 + 0 = 1$
    Answer: 1 Kings 10:1

21. Book: $13 + 14 = 27$
    Chapter: $0 + 1 = 1$
    Verse: $16 + 4 = 20$
    Answer: Daniel 1:20

22. Book: $9 + 4 = 13$
    Chapter: $15 + 2 = 17$
    Verse: $1 + 1 = 2$
    Answer: 1 Chronicles 17:2

23. Book: $22 + 17 = 39$
    Chapter: $3 + 0 = 3$
    Verse: $0 + 1 = 1$
    Answer: Malachi 3:1

24. Book: $13 + 13 = 26$
    Chapter: $29 + 7 = 36$
    Verse: $13 + 14 = 27$
    Answer: Ezekiel 36:27

25. Book: $17 + 3 = 20$
    Chapter: $13 + 9 = 22$
    Verse: $1 + 0 = 1$
    Answer: Proverbs 22:1

26. Book: $9 + 5 = 14$
    Chapter: $8 + 2 = 10$
    Verse: $11 + 3 = 14$
    Answer: 2 Chronicles 10:14

27. Book: $20 + 8 = 28$
    Chapter: $10 + 2 = 12$
    Verse: $5 + 1 = 6$
    Answer: Hosea 12:6

28. Book: $30 + 2 = 32$
    Chapter: $1 + 3 = 4$
    Verse: $10 + 1 = 11$
    Answer: Jonah 4:11

29. Book: $29 + 8 = 37$
    Chapter: $0 + 1 = 1$
    Verse: $5 + 0 = 5$
    Answer: Haggai 1:5

30. Book: $25 + 5 = 30$
    Chapter: $2 + 1 = 3$
    Verse: $0 + 3 = 3$
    Answer: Amos 3:3

31. Book: $28 + 7 = 35$
    Chapter: $2 + 0 = 2$
    Verse: $8 + 7 = 15$
    Answer: Habakkuk 2:15

32. Book: $20 + 13 = 33$
    Chapter: $3 + 4 = 7$
    Verse: $5 + 2 = 7$
    Answer: Micah 7:7

33. Book: $2 + 12 = 14$
    Chapter: $9 + 8 = 17$
    Verse: $18 + 2 = 20$
    Answer: 2 Chronicles 17:20

34. Book: $10 + 11 = 21$
    Chapter: $1 + 6 = 7$
    Verse: $5 + 4 = 9$
    Answer: Ecclesiastes 7:9

35. Book: $14 + 1 = 15$
    Chapter: $2 + 4 = 6$
    Verse: $7 + 5 = 12$
    Answer: Ezra 6:12

36. Book: $16 + 6 = 22$
    Chapter: $1 + 2 = 3$
    Verse: $10 + 1 = 11$
    Answer: Song of Solomon 3:11

37. Book: $30 + 6 = 36$
    Chapter: $1 + 1 = 2$
    Verse: $14 + 1 = 15$
    Answer: Zephaniah 2:15

38. Book: $20 + 9 = 29$
    Chapter: $0 + 1 = 1$
    Verse: $7 + 7 = 14$
    Answer: Joel 1:14

39. Book: $20 + 5 = 25$
    Chapter: $4 + 1 = 5$
    Verse: $14 + 5 = 19$
    Answer: Lamentations 5:19

40. Book: $18 + 1 = 19$
    Chapter: $7 + 2 = 9$
    Verse: $1 + 0 = 1$
    Answer: Psalm 19:1

41. Book: $19 + 4 = 23$
    Chapter: $20 + 38 = 58$
    Verse: $4 + 2 = 6$
    Answer: Isaiah 58:6

42. Book: $3 + 2 = 5$
    Chapter: $7 + 0 = 7$
    Verse: $11 + 1 = 12$
    Answer: Deuteronomy 7:12

43. Book: $16 + 4 = 20$
    Chapter: $20 + 9 = 29$
    Verse: $10 + 4 = 14$
    Answer: Proverbs 29:14

44. Book: $9 + 9 = 18$
    Chapter: $14 + 14 = 28$
    Verse: $19 + 9 = 28$
    Answer: Job 28:28

45. Book: $12 + 14 = 26$
    Chapter: $22 + 22 = 44$
    Verse: $15 + 8 = 23$
    Answer: Ezekiel 44:23

46. Book: $12 + 12 = 24$
    Chapter: $21 + 4 = 25$
    Verse: $10 + 7 = 17$
    Answer: Jeremiah 25:17

47. Book: $1 + 1 = 2$
    Chapter: $30 + 2 = 32$
    Verse: $6 + 6 = 12$
    Answer: Exodus 32:12

48. Book: $20 + 4 = 24$
    Chapter: $25 + 25 = 50$
    Verse: $24 + 10 = 34$
    Answer: Jeremiah 50:34

49. Book: $2 + 2 = 4$
    Chapter: $20 + 2 = 22$
    Verse: $14 + 14 = 28$
    Answer: Numbers 22:28

50. Book: $19 + 6 = 25$
    Chapter: $0 + 1 = 1$
    Verse: $19 + 1 = 20$
    Answer: Lamentations 1:20

SUBTRACTIONS
Old Testament Answers

1. Book: $39 - 20 = 19$
   Chapter: $40 - 6 = 34$
   Verse: $2 - 1 = 1$
   Answer: Psalm 34:1

2. Book: $6 - 5 = 1$
   Chapter: $8 - 6 = 2$
   Verse: $3 - 1 = 2$
   Answer: Genesis 2:2

3. Book: $30 - 7 = 23$
   Chapter: $50 - 13 = 37$
   Verse: $32 - 12 = 20$
   Answer: Isaiah 37:20

4. Book: $11 - 8 = 3$
   Chapter: $29 - 10 = 19$
   Verse: $50 - 32 = 20$
   Answer: Leviticus 19:20

5. Book: $10 - 3 = 7$
   Chapter: $50 - 30 = 20$
   Verse: $42 - 14 = 28$
   Answer: Judges 20:28

6. Book: $18 - 13 = 5$
   Chapter: $20 - 6 = 14$
   Verse: $12 - 4 = 8$
   Answer: Deuteronomy 14:8

7. Book: $20 - 2 = 18$
   Chapter: $50 - 8 = 42$
   Verse: $15 - 3 = 12$
   Answer: Job 42:12

8. Book: $26 - 2 = 24$
   Chapter: $15 - 8 = 7$
   Verse: $21 - 10 = 11$
   Answer: Jeremiah 7:11

9. Book: $25 - 12 = 13$
   Chapter: $16 - 5 = 11$
   Verse: $2 - 1 = 1$
   Answer: 1 Chronicles 11:1

10. Book: $29 - 3 = 26$
    Chapter: $30 - 10 = 20$
    Verse: $43 - 16 = 27$
    Answer: Ezekiel 20:27

11. Book: $50 - 11 = 39$
    Chapter: $6 - 3 = 3$
    Verse: $14 - 8 = 6$
    Answer: Malachi 3:6

12. Book: $26 - 5 = 21$
    Chapter: $23 - 11 = 12$
    Verse: $36 - 23 = 13$
    Answer: Ecclesiastes 12:13

13. Book: $13 - 4 = 9$
    Chapter: $31 - 13 = 18$
    Verse: $25 - 11 = 14$
    Answer: 1 Samuel 18:14

14. Book: $55 - 35 = 20$
    Chapter: $35 - 15 = 20$
    Verse: $23 - 1 = 22$
    Answer: Proverbs 20:22

15. Book: 39 – 24 = 15
    Chapter: 12 – 9 = 3
    Verse: 18 – 7 = 11
    Answer: Ezra 3:11

16. Book: 39 – 25 = 14
    Chapter: 20 – 6 = 14
    Verse: 50 – 21 = 29
    Answer: 2 Chronicles 14:29

17. Book: 39 – 12 = 27
    Chapter: 10 – 4 = 6
    Verse: 28 – 1 = 27
    Answer: Daniel 6:27

18. Book: 39 – 35 = 4
    Chapter: 14 – 3 = 11
    Verse: 23 – 8 = 15
    Answer: Numbers 11:15

19. Book: 39 – 1 = 38
    Chapter: 14 – 4 = 10
    Verse: 20 – 8 = 12
    Answer: Zechariah 10:12

20. Book: 26 – 19 = 7
    Chapter: 40 – 24 = 16
    Verse: 32 – 14 = 18
    Answer: Judges 16:18

21. Book: 36 – 11 = 25
    Chapter: 42 – 39 = 3
    Verse: 75 – 11 = 64
    Answer: Lamentations 3:64

22. Book: 39 – 7 = 32
    Chapter: 2 – 1 = 1
    Verse: 10 – 7 = 3
    Answer: Jonah 1:3

23. Book: 39 – 33 = 6
    Chapter: 15 – 5 = 10
    Verse: 18 – 6 = 12
    Answer: Joshua 10:12

24. Book: 39 – 29 = 10
    Chapter: 26 – 4 = 22
    Verse: 50 – 3 = 47
    Answer: 2 Samuel 22:47

25. Book: 20 – 4 = 16
    Chapter: 14 – 13 = 1
    Verse: 22 – 11 = 11
    Answer: Nehemiah 1:11

26. Book: 39 – 27 = 12
    Chapter: 27 – 4 = 23
    Verse: 8 – 5 = 3
    Answer: 2 Kings 23:3

27. Book: 39 – 9 = 30
    Chapter: 18 – 13 = 5
    Verse: 30 – 6 = 24
    Answer: Amos 5:24

28. Book: 39 – 22 = 17
    Chapter: 19 – 16 = 3
    Verse: 7 – 1 = 6
    Answer: Esther 3:6

29. Book: 39 – 6 = 33
    Chapter: 18 – 16 = 2
    Verse: 4 – 3 = 1
    Answer: Micah 2:1

30. Book: 20 – 10 = 10
    Chapter: 23 – 11 = 12
    Verse: 5 – 1 = 4
    Answer: 2 Samuel 12:4

MULTIPLICATIONS
Answers Old Testament

1. Book: $2 \times 19 = 38$
   Chapter: $2 \times 1 = 2$
   Verse: $4 \times 2 = 8$
   Answer   Zechariah 2:8

2. Book: $23 \times 1 = 23$
   Chapter: $5 \times 1 = 5$
   Verse: $3 \times 7 = 21$
   Answer: Isaiah 5:21

3. Book: $7 \times 2 = 14$
   Chapter: $1 \times 1 = 1$
   Verse: $5 \times 2 = 10$
   Answer: 2 Chronicles 1:10

4. Book: $19 \times 1 = 19$
   Chapter: $49 \times 2 = 98$
   Verse: $2 \times 2 = 4$
   Answer: Psalm 98:4

5. Book: $1 \times 4 = 4$
   Chapter: $6 \times 3 = 18$
   Verse: $3 \times 7 = 21$
   Answer: Numbers 18:21

6. Book: $23 \times 1 = 23$
   Chapter: $3 \times 3 = 9$
   Verse: $2 \times 3 = 6$
   Answer: Isaiah 9:6

7. Book: $3 \times 1 = 3$
   Chapter: $6 \times 4 = 24$
   Verse: $2 \times 1 = 2$
   Answer: Leviticus 24:2

8. Book: $8 \times 4 = 32$
   Chapter: $2 \times 1 = 2$
   Verse: $5 \times 2 = 10$
   Answer: Jonah 2:10

9. Book: $3 \times 2 = 6$
   Chapter: $2 \times 1 = 2$
   Verse: $3 \times 3 = 9$
   Answer: Joshua 2:9

10. Book: $1 \times 1 = 1$
    Chapter: $1 \times 1 = 1$
    Verse: $29 \times 1 = 29$
    Answer: Genesis 1:29

11. Book: $2 \times 12 = 24$
    Chapter: $5 \times 2 = 10$
    Verse: $4 \times 3 = 12$
    Answer: Jeremiah 10:12

12. Book: $3 \times 7 = 21$
    Chapter: $2 \times 2 = 4$
    Verse: $3 \times 2 = 6$
    Answer: Ecclesiastes 4:6

13. Book: $4 \times 3 = 12$
    Chapter: $1 \times 2 = 2$
    Verse: $11 \times 1 = 11$
    Answer: 2 Kings 2:11

14. Book: $1 \times 19 = 19$
    Chapter: $5 \times 5 = 25$
    Verse: $4 \times 5 = 20$
    Answer: Psalm 25:20

15. Book: $27 \times 1 = 27$
    Chapter: $1 \times 2 = 2$
    Verse: $23 \times 1 = 23$
    Answer: Daniel 2:23

16. Book: $5 \times 2 = 10$
    Chapter: $2 \times 2 = 4$
    Verse: $2 \times 2 = 4$
    Answer: 2 Samuel 4:4

17. Book: $5 \times 7 = 35$
    Chapter: $2 \times 1 = 2$
    Verse: $2 \times 2 = 4$
    Answer: Habakkuk 2:4

18. Book: $1 \times 2 = 2$
    Chapter: $2 \times 1 = 2$
    Verse: $5 \times 2 = 10$
    Answer: Exodus 2:10

19. Book: $2 \times 2 = 4$
    Chapter: $13 \times 2 = 26$
    Verse: $3 \times 18 = 54$
    Answer: Numbers 26:54

20. Book: $7 \times 4 = 28$
    Chapter: $2 \times 3 = 6$
    Verse: $3 \times 1 = 3$
    Answer: Hosea 6:3

21. Book: $2 \times 17 = 34$
    Chapter: $2 \times 1 = 2$
    Verse: $4 \times 2 = 8$
    Answer: Nahum 2:8

22. Book: $2 \times 10 = 20$
    Chapter: $1 \times 3 = 3$
    Verse: $13 \times 1 = 13$
    Answer: Proverbs 3:13

23. Book: $5 \times 1 = 5$
    Chapter: $2 \times 17 = 34$
    Verse: $1 \times 7 = 7$
    Answer: Deuteronomy 34:7

24. Book: $2 \times 13 = 26$
    Chapter: $43 \times 1 = 43$
    Verse: $1 \times 5 = 5$
    Answer: Ezekiel 43:5

25. Book: $1 \times 7 = 7$
    Chapter: $5 \times 1 = 5$
    Verse: $2 \times 1 = 2$
    Answer: Judges 5:2

26. Book: $1 \times 29 = 29$
    Chapter: $2 \times 1 = 2$
    Verse: $1 \times 13 = 13$
    Answer: Joel 2:13

27. Book: $6 \times 3 = 18$
    Chapter: $2 \times 19 = 38$
    Verse: $2 \times 2 = 4$
    Answer: Job 38:4

28. Book: $11 \times 1 = 11$
    Chapter: $2 \times 11 = 22$
    Verse: $1 \times 23 = 23$
    Answer: 1 Kings 22:23

29. Book: $3 \times 3 = 9$
    Chapter: $6 \times 5 = 30$
    Verse: $5 \times 2 = 10$
    Answer: 1 Samuel 30:10

30. Book: $4 \times 2 = 8$
    Chapter: $1 \times 2 = 2$
    Verse: $3 \times 6 = 18$
    Answer: Ruth 2:18

DIVISIONS
Old Testament Answers

1. Book: $230 \div 10 = 23$
   Chapter: $52 \div 2 = 26$
   Verse: $9 \div 3 = 3$
   Answer: Isaiah 26:3

2. Book: $380 \div 10 = 38$
   Chapter: $36 \div 3 = 12$
   Verse: $24 \div 3 = 8$
   Answer: Zechariah 12:8

3. Book: $42 \div 3 = 14$
   Chapter: $30 \div 5 = 6$
   Verse: $32 \div 2 = 16$
   Answer: 2 Chronicles 6:16

4. Book: $100 \div 25 = 4$
   Chapter: $70 \div 5 = 14$
   Verse: $96 \div 12 = 8$
   Answer: Numbers 14:8

5. Book: $96 \div 4 = 24$
   Chapter: $87 \div 3 = 29$
   Verse: $33 \div 3 = 11$
   Answer: Jeremiah 29:11

6. Book: $42 \div 14 = 3$
   Chapter: $36 \div 2 = 18$
   Verse: $9 \div 3 = 3$
   Answer: Leviticus 18:3

7. Book: $96 \div 3 = 32$
   Chapter: $8 \div 8 = 1$
   Verse: $34 \div 2 = 17$
   Answer: Jonah 1:17

8. Book: $42 \div 7 = 6$
   Chapter: $27 \div 9 = 3$
   Verse: $28 \div 7 = 4$
   Answer: Joshua 3:4

9. Book: $42 \div 2 = 21$
   Chapter: $15 \div 3 = 5$
   Verse: $4 \div 4 = 1$
   Answer: Ecclesiastes 5:1

10. Book: $3 \div 3 = 1$
    Chapter: $21 \div 7 = 3$
    Verse: $12 \div 2 = 6$
    Answer: Genesis 3:6

11. Book: $24 \div 2 = 12$
    Chapter: $25 \div 5 = 5$
    Verse: $52 \div 2 = 26$
    Answer: 2 Kings 5:26

12. Book: $80 \div 4 = 20$
    Chapter: $96 \div 6 = 16$
    Verse: $180 \div 20 = 9$
    Answer: Proverbs 16:9

13. Book: $54 \div 2 = 27$
    Chapter: $45 \div 9 = 5$
    Verse: $51 \div 3 = 17$
    Answer: Daniel 5:17

14. Book: $40 \div 4 = 10$
    Chapter: $30 \div 5 = 6$
    Verse: $36 \div 3 = 12$
    Answer: 2 Samuel 6:12

15. Book: $350 \div 10 = 35$
    Chapter: $45 \div 15 = 3$
    Verse: $60 \div 6 = 10$
    Answer: Habakkuk 3:10

16. Book: $30 \div 15 = 2$
    Chapter: $16 \div 4 = 4$
    Verse: $15 \div 5 = 3$
    Answer: Exodus 4:3

17. Book: $32 \div 4 = 8$
    Chapter: $1 \div 1 = 1$
    Verse: $160 \div 10 = 16$
    Answer: Ruth 1:16

18. Book: $56 \div 2 = 28$
    Chapter: $42 \div 3 = 14$
    Verse: $7 \div 7 = 1$
    Answer: Hosea 14:1

19. Book: $102 \div 3 = 34$
    Chapter: $30 \div 10 = 3$
    Verse: $42 \div 6 = 7$
    Answer: Nahum 3:7

20. Book: $60 \div 12 = 5$
    Chapter: $124 \div 4 = 31$
    Verse: $84 \div 7 = 12$
    Answer: Deuteronomy 31:12

21. Book: $52 \div 2 = 26$
    Chapter: $39 \div 3 = 13$
    Verse: $21 \div 7 = 3$
    Answer: Ezekiel 13:3

22. Book: $49 \div 7 = 7$
    Chapter: $44 \div 4 = 11$
    Verse: $30 \div 6 = 5$
    Answer: Judges 11:5

23. Book: $87 \div 3 = 29$
    Chapter: $18 \div 9 = 2$
    Verse: $46 \div 2 = 23$
    Answer: Joel 2:23

24. Book: $36 \div 2 = 18$
    Chapter: $10 \div 5 = 2$
    Verse: $6 \div 6 = 1$
    Answer: Job 2:1

25. Book: $44 \div 4 = 11$
    Chapter: $66 \div 3 = 22$
    Verse: $86 \div 2 = 43$
    Answer: 1 Kings 22:43

26. Book: $45 \div 5 = 9$
    Chapter: $81 \div 3 = 27$
    Verse: $11 \div 11 = 1$
    Answer: 1 Samuel 27:1

27. Book: $48 \div 3 = 16$
    Chapter: $20 \div 5 = 4$
    Verse: $9 \div 9 = 1$
    Answer: Nehemiah 4:1

28. Book: $120 \div 4 = 30$
    Chapter: $40 \div 5 = 8$
    Verse: $77 \div 7 = 11$
    Answer: Amos 8:11

29. Book: $80 \div 5 = 16$
    Chapter: $54 \div 9 = 6$
    Verse: $160 \div 10 = 16$
    Answer: Nehemiah 6:16

30. Book: $36 \div 3 = 12$
    Chapter: $40 \div 10 = 4$
    Verse: $35 \div 5 = 7$
    Answer: 2 Kings 4:7

ADDITIONS
New Testament Answers

1. Book: $2 + 0 = 2$
   Chapter: $9 + 4 = 13$
   Verse: $17 + 18 = 35$
   Answer: Mark 13:35

2. Book: $3 + 1 = 4$
   Chapter: $11 + 3 = 14$
   Verse: $2 + 1 = 3$
   Answer: John 14:3

3. Book: $2 + 4 = 6$
   Chapter: $2 + 3 = 5$
   Verse: $1 + 4 = 5$
   Answer: Romans 5:5

4. Book: $3 + 5 = 8$
   Chapter: $1 + 4 = 5$
   Verse: $8 + 2 = 10$
   Answer: 2 Corinthians 5:10

5. Book: $1 + 0 = 1$
   Chapter: $3 + 4 = 7$
   Verse: $0 + 1 = 1$
   Answer: Matthew 7:1

6. Book: $7 + 2 = 9$
   Chapter: $2 + 3 = 5$
   Verse: $8 + 7 = 15$
   Answer: Galatians 5:15

7. Book: $12 + 1 = 13$
   Chapter: $5 + 0 = 5$
   Verse: $10 + 12 = 22$
   Answer: 1 Thessalonians 5:22

8. Book: $8 + 5 = 13$
   Chapter: $1 + 0 = 1$
   Verse: $7 + 2 = 9$
   Answer: 1 Thessalonians 1:9

9. Book: $3 + 0 = 3$
   Chapter: $0 + 1 = 1$
   Verse: $20 + 22 = 42$
   Answer: Luke 1:42

10. Book: $3 + 2 = 5$
    Chapter: $1 + 4 = 5$
    Verse: $17 + 2 = 19$
    Answer: Acts 5:19

11. Book: $6 + 5 = 11$
    Chapter: $1 + 0 = 1$
    Verse: $7 + 4 = 11$
    Answer: Philippians 1:11

12. Book: $1 + 13 = 14$
    Chapter: $0 + 1 = 1$
    Verse: $2 + 2 = 4$
    Answer: 2 Thessalonians 1:4

13. Book: $2 + 14 = 16$
    Chapter: $1 + 1 = 2$
    Verse: $14 + 2 = 16$
    Answer: 2 Timothy 2:16

14. Book: $9 + 1 = 10$
    Chapter: $1 + 1 = 2$
    Verse: $5 + 3 = 8$
    Answer: Ephesians 2:8

15. Book: $8 + 9 = 17$
    Chapter: $0 + 1 = 1$
    Verse: $6 + 9 = 15$
    Answer: Titus 1:15

16. Book: 4 + 15 = 19
    Chapter: 6 + 7 = 13
    Verse: 1 + 1 = 2
    Answer: Hebrews 13:2

17. Book: 17 + 4 = 21
    Chapter: 0 + 1 = 1
    Verse: 16 + 6 = 22
    Answer: 1 Peter 1:22

18. Book: 2 + 0 = 2
    Chapter: 3 + 3 = 6
    Verse: 27 + 6 = 33
    Answer: Mark 6:33

19. Book: 15 + 3 = 18
    Chapter: 0 + 1 = 1
    Verse: 7 + 11 = 18
    Answer: Philemon 1:18

20. Book: 10 + 12 = 22
    Chapter: 0 + 1 = 1
    Verse: 16 + 4 = 20
    Answer: 2 Peter 1:20

21. Book: 16 + 7 = 23
    Chapter: 0 + 1 = 1
    Verse: 5 + 5 = 10
    Answer: 1 John 1:10

22. Book: 15 + 12 = 27
    Chapter: 2 + 1 = 3
    Verse: 11 + 9 = 20
    Answer: Revelation 3:20

23. Book: 0 + 1 = 1
    Chapter: 3 + 2 = 5
    Verse: 6 + 1 = 7
    Answer: Matthew 5:7

24. Book: 6 + 6 = 12
    Chapter: 1 + 0 = 1
    Verse: 6 + 3 = 9
    Answer: Colossians 1:9

25. Book: 16 + 4 = 20
    Chapter: 1 + 0 = 1
    Verse: 17 + 5 = 22
    Answer: James 1:22

26. Book: 18 + 6 = 24
    Chapter: 0 + 1 = 1
    Verse: 3 + 3 = 6
    Answer: 2 John 1:6

27. Book: 8 + 6 = 14
    Chapter: 2 + 1 = 3
    Verse: 9 + 7 = 16
    Answer: 2 Thessalonians 3:16

28. Book: 18 + 1 = 19
    Chapter: 4 + 2 = 6
    Verse: 10 + 2 = 12
    Answer: Hebrews 6:12

29. Book: 5 + 10 = 15
    Chapter: 0 + 1 = 1
    Verse: 5 + 3 = 8
    Answer: 1 Timothy 1:8

30. Book: 6 + 1 = 7
    Chapter: 1 + 1 = 2
    Verse: 9 + 3 = 12
    Answer: 1 Corinthians 2:12

SUBTRACTIONS
New Testament Answers

1. Book: $9 - 5 = 4$
   Chapter: $6 - 2 = 4$
   Verse: $50 - 8 = 42$
   Answer: John 4:42

2. Book: $25 - 19 = 6$
   Chapter: $10 - 3 = 7$
   Verse: $19 - 7 = 12$
   Answer: Romans 7:12

3. Book: $15 - 6 = 9$
   Chapter: $8 - 2 = 6$
   Verse: $22 - 15 = 7$
   Answer: Galatians 6:7

4. Book: $6 - 5 = 1$
   Chapter: $27 - 13 = 14$
   Verse: $36 - 22 = 14$
   Answer: Matthew 14:14

5. Book: $17 - 12 = 5$
   Chapter: $20 - 15 = 5$
   Verse: $30 - 1 = 29$
   Answer: Acts 5:29

6. Book: $50 - 42 = 8$
   Chapter: $20 - 12 = 8$
   Verse: $24 - 10 = 14$
   Answer: 2 Corinthians 8:14

7. Book: $16 - 5 = 11$
   Chapter: $6 - 4 = 2$
   Verse: $5 - 3 = 2$
   Answer: Philippians 2:2

8. Book: $16 - 13 = 3$
   Chapter: $17 - 11 = 6$
   Verse: $36 - 9 = 27$
   Answer: Luke 6:27

9. Book: $20 - 8 = 12$
   Chapter: $5 - 2 = 3$
   Verse: $7 - 5 = 2$
   Answer: Colossians 3:2

10. Book: $20 - 4 = 16$
    Chapter: $16 - 15 = 1$
    Verse: $18 - 11 = 7$
    Answer: 2 Timothy 1:7

11. Book: $26 - 16 = 10$
    Chapter: $10 - 6 = 4$
    Verse: $32 - 3 = 29$
    Answer: Ephesians 4:29

12. Book: $20 - 3 = 17$
    Chapter: $6 - 4 = 2$
    Verse: $8 - 2 = 6$
    Answer: Titus 2:6

13. Book: $3 - 2 = 1$
    Chapter: $20 - 13 = 7$
    Verse: $29 - 11 = 18$
    Answer: Matthew 7:18

14. Book: $17 - 4 = 13$
    Chapter: $3 - 1 = 2$
    Verse: $20 - 7 = 13$
    Answer: 1 Thessalonians 2:13

15. Book: $24 - 9 = 15$
    Chapter: $2 - 1 = 1$
    Verse: $36 - 24 = 12$
    Answer: 1 Timothy 1:12

16. Book: 27 – 8 = 19
    Chapter: 33 – 21 = 12
    Verse: 18 – 17 = 1
    Answer: Hebrews 12:1

17. Book: 27 – 6 = 21
    Chapter: 12 – 10 = 2
    Verse: 16 – 7 = 9
    Answer: 1 Peter 2:9

18. Book: 35 – 12 = 23
    Chapter: 4 – 2 = 2
    Verse: 10 – 6 = 4
    Answer: 1 John 2:4

19. Book: 16 – 14 = 2
    Chapter: 20 – 13 = 7
    Verse: 16 – 7 = 9
    Answer: Mark 7:9

20. Book: 18 – 4 = 14
    Chapter: 7 – 6 = 1
    Verse: 16 – 5 = 11
    Answer: 2 Thessalonians 1:11

21. Book: 27 – 7 = 20
    Chapter: 2 – 1 = 1
    Verse: 40 – 23 = 17
    Answer: James 1:17

22. Book: 64 – 42 = 22
    Chapter: 7 – 5 = 2
    Verse: 40 – 18 = 22
    Answer: 2 Peter 2:22

23. Book: 66 – 39 = 27
    Chapter: 15 – 12 = 3
    Verse: 31 – 10 = 21
    Answer: Revelation 3:21

24. Book: 45 – 21 = 24
    Chapter: 2 – 1 = 1
    Verse: 8 – 6 = 2
    Answer: 2 John 1:22

25. Book: 16 – 10 = 6
    Chapter: 20 – 7 = 13
    Verse: 19 – 6 = 13
    Answer: Romans 13:13

26. Book: 27 – 9 = 18
    Chapter: 5 – 4 = 1
    Verse: 30 – 5 = 25
    Answer: Philemon 1:25

27. Book: 13 – 3 = 10
    Chapter: 10 – 6 = 4
    Verse: 30 – 4 = 26
    Answer: Ephesians 4:26

28. Book: 20 – 3 = 17
    Chapter: 8 – 6 = 2
    Verse: 15 – 2 = 13
    Answer: Titus 2:13

29. Book: 3 – 1 = 2
    Chapter: 14 – 3 = 11
    Verse: 30 – 6 = 24
    Answer: Mark 11:24

30. Book: 25 – 18 = 7
    Chapter: 16 – 3 = 13
    Verse: 1 – 0 = 1
    Answer: 1 Corinthians 13:1

MULTIPLICATIONS
New Testament Answers

1. Book: $2 \times 1 = 2$
    Chapter: $3 \times 4 = 12$
    Verse: $1 \times 1 = 1$
    Answer: Mark 12:1

2. Book: $5 \times 1 = 5$
    Chapter: $1 \times 2 = 2$
    Verse: $5 \times 9 = 45$
    Answer: Acts 2:45

3. Book: $7 \times 1 = 7$
    Chapter: $2 \times 7 = 14$
    Verse: $5 \times 8 = 40$
    Answer: 1 Corinthians 14:40

4. Book: $1 \times 11 = 11$
    Chapter: $4 \times 1 = 4$
    Verse: $13 \times 1 = 13$
    Answer: Philippians 4:13

5. Book: $3 \times 3 = 9$
    Chapter: $2 \times 2 = 4$
    Verse: $5 \times 4 = 20$
    Answer: Galatians 4:20

6. Book: $13 \times 1 = 13$
    Chapter: $1 \times 3 = 3$
    Verse: $3 \times 3 = 9$
    Answer: 1 Thessalonians 3:9

7. Book: $5 \times 3 = 15$
    Chapter: $1 \times 3 = 3$
    Verse: $2 \times 1 = 2$
    Answer: 1 Timothy 3:2

8. Book: $4 \times 4 = 16$
    Chapter: $2 \times 1 = 2$
    Verse: $4 \times 2 = 8$
    Answer: 2 Timothy 2:8

9. Book: $3 \times 1 = 3$
    Chapter: $2 \times 1 = 2$
    Verse: $5 \times 8 = 40$
    Answer: Luke 2:40

10. Book: $3 \times 2 = 6$
    Chapter: $3 \times 3 = 9$
    Verse: $4 \times 4 = 16$
    Answer: Romans 9:16

11. Book: $1 \times 1 = 1$
    Chapter: $6 \times 2 = 12$
    Verse: $5 \times 7 = 35$
    Answer: Matthew 12:35

12. Book: $2 \times 5 = 10$
    Chapter: $2 \times 2 = 4$
    Verse: $5 \times 3 = 15$
    Answer: Ephesians 4:15

13. Book: $3 \times 5 = 15$
    Chapter: $2 \times 2 = 4$
    Verse: $3 \times 4 = 12$
    Answer: 1 Timothy 4:12

14. Book: $2 \times 2 = 4$
    Chapter: $4 \times 2 = 8$
    Verse: $7 \times 1 = 7$
    Answer: John 8:7

15. Book: $2 \times 4 = 8$
    Chapter: $2 \times 2 = 4$
    Verse: $1 \times 2 = 2$
    Answer: 2 Corinthians 4:2

16. Book: $11 \times 1 = 11$
Chapter: $2 \times 1 = 2$
Verse: $7 \times 2 = 14$
Answer: Philippians 2:14

17. Book: $2 \times 7 = 14$
Chapter: $1 \times 3 = 3$
Verse: $5 \times 2 = 10$
Answer: 2 Thessalonians 3:10

18. Book: $17 \times 1 = 17$
Chapter: $3 \times 1 = 3$
Verse: $2 \times 4 = 8$
Answer: Titus 3:8

19. Book: $11 \times 1 = 11$
Chapter: $2 \times 2 = 4$
Verse: $4 \times 2 = 8$
Answer: Philippians 4:8

20. Book: $19 \times 1 = 19$
Chapter: $11 \times 1 = 11$
Verse: $1 \times 1 = 1$
Answer: Hebrews 11:1

21. Book: $3 \times 7 = 21$
Chapter: $1 \times 3 = 3$
Verse: $4 \times 2 = 8$
Answer: 1 Peter 3:8

22. Book: $13 \times 1 = 13$
Chapter: $2 \times 2 = 4$
Verse: $4 \times 4 = 16$
Answer: 1 Thessalonians 4:16

23. Book: $2 \times 9 = 18$
Chapter: $1 \times 1 = 1$
Verse: $2 \times 2 = 4$
Answer: Philemon 1:4

24. Book: $2 \times 10 = 20$
Chapter: $1 \times 1 = 1$
Verse: $4 \times 2 = 8$
Answer: James 1:8

25. Book: $3 \times 9 = 27$
Chapter: $2 \times 11 = 22$
Verse: $4 \times 3 = 12$
Answer: Revelation 22:12

26. Book: $3 \times 4 = 12$
Chapter: $1 \times 1 = 1$
Verse: $3 \times 1 = 3$
Answer: Colossians 1:3

27. Book: $2 \times 1 = 2$
Chapter: $4 \times 2 = 8$
Verse: $4 \times 10 = 40$
Answer: Mark 8:40

28. Book: $2 \times 7 = 14$
Chapter: $1 \times 2 = 2$
Verse: $13 \times 1 = 13$
Answer: 2 Thessalonians 2:13

29. Book: $10 \times 1 = 10$
Chapter: $1 \times 6 = 6$
Verse: $2 \times 5 = 10$
Answer: Ephesians 6:10

30. Book: $1 \times 1 = 1$
Chapter: $3 \times 3 = 9$
Verse: $6 \times 3 = 18$
Answer: Matthew 9:18

DIVISIONS
New Testament Answers

1. Book: $16 \div 4 = 4$
   Chapter: $48 \div 6 = 8$
   Verse: $160 \div 5 = 32$
   Answer: John 8:32

2. Book: $42 \div 7 = 6$
   Chapter: $56 \div 7 = 8$
   Verse: $42 \div 3 = 14$
   Answer: Romans 8:14

3. Book: $26 \div 13 = 2$
   Chapter: $18 \div 6 = 3$
   Verse: $80 \div 20 = 4$
   Answer: Mark 3:4

4. Book: $14 \div 2 = 7$
   Chapter: $54 \div 6 = 9$
   Verse: $26 \div 2 = 13$
   Answer: 1 Corinthians 9:13

5. Book: $45 \div 5 = 9$
   Chapter: $60 \div 10 = 6$
   Verse: $9 \div 3 = 3$
   Answer: Galatians 6:3

6. Book: $144 \div 12 = 12$
   Chapter: $45 \div 15 = 3$
   Verse: $75 \div 3 = 25$
   Answer: Colossians 3:25

7. Book: $16 \div 1 = 16$
   Chapter: $2 \div 1 = 2$
   Verse: $95 \div 5 = 19$
   Answer: 2 Timothy 2:19

8. Book: $9 \div 3 = 3$
   Chapter: $48 \div 8 = 6$
   Verse: $90 \div 2 = 45$
   Answer: Luke 6:45

9. Book: $39 \div 3 = 13$
   Chapter: $60 \div 12 = 5$
   Verse: $85 \div 5 = 17$
   Answer: I Thessalonians 5:17

10. Book: $55 \div 11 = 5$
    Chapter: $63 \div 7 = 9$
    Verse: $72 \div 12 = 6$
    Answer: Acts 9:6

11. Book: $1 \div 1 = 1$
    Chapter: $32 \div 2 = 16$
    Verse: $96 \div 6 = 16$
    Answer: Matthew 16:16

12. Book: $40 \div 5 = 8$
    Chapter: $26 \div 2 = 13$
    Verse: $35 \div 5 = 7$
    Answer: 2 Corinthians 13:7

13. Book: $270 \div 27 = 10$
    Chapter: $50 \div 10 = 5$
    Verse: $60 \div 3 = 20$
    Answer: Ephesians 5:20

14. Book: $14 \div 1 = 14$
    Chapter: $36 \div 12 = 3$
    Verse: $24 \div 4 = 6$
    Answer: 2 Thessalonians 3:6

15. Book: $36 \div 2 = 18$
    Chapter: $1 \div 1 = 1$
    Verse: $10 \div 2 = 5$
    Answer: Philemon 1:5

16. Book: $20 \div 1 = 20$
    Chapter: $10 \div 10 = 1$
    Verse: $57 \div 3 = 19$
    Answer: James 1:19

17. Book: $42 \div 2 = 21$
    Chapter: $12 \div 3 = 4$
    Verse: $56 \div 7 = 8$
    Answer: 1 Peter 4:8

18. Book: $46 \div 2 = 23$
    Chapter: $23 \div 23 = 1$
    Verse: $28 \div 4 = 7$
    Answer: 1 John 1:7

19. Book: $42 \div 3 = 14$
    Chapter: $26 \div 13 = 2$
    Verse: $12 \div 4 = 3$
    Answer: 2 Thessalonians 2:3

20. Book: $6 \div 3 = 2$
    Chapter: $48 \div 3 = 16$
    Verse: $54 \div 9 = 6$
    Answer: Mark 16:6

21. Book: $60 \div 4 = 15$
    Chapter: $100 \div 50 = 2$
    Verse: $80 \div 10 = 8$
    Answer: 1 Timothy 2:8

22. Book: $19 \div 1 = 19$
    Chapter: $8 \div 2 = 4$
    Verse: $40 \div 10 = 4$
    Answer: Hebrews 4:4

23. Book: $135 \div 5 = 27$
    Chapter: $84 \div 6 = 14$
    Verse: $48 \div 4 = 12$
    Answer: Revelation 14:12

24. Book: $49 \div 7 = 7$
    Chapter: $60 \div 15 = 4$
    Verse: $28 \div 2 = 14$
    Answer: 1 Corinthians 4:14

25. Book: $15 \div 3 = 5$
    Chapter: $34 \div 2 = 17$
    Verse: $90 \div 3 = 30$
    Answer: Acts 17:30

26. Book: $66 \div 6 = 11$
    Chapter: $4 \div 2 = 2$
    Verse: $44 \div 4 = 11$
    Answer: Philippians 2:11

27. Book: $108 \div 4 = 27$
    Chapter: $147 \div 7 = 21$
    Verse: $20 \div 5 = 4$
    Answer: Revelation 21:4

28. Book: $1 \div 1 = 1$
    Chapter: $100 \div 10 = 10$
    Verse: $90 \div 3 = 30$
    Answer: Matthew 10:30

29. Book: $44 \div 2 = 22$
    Chapter: $30 \div 10 = 3$
    Verse: $70 \div 5 = 14$
    Answer: 2 Peter 3:14

30. Book: $70 \div 7 = 10$
    Chapter: $44 \div 11 = 4$
    Verse: $96 \div 3 = 32$
    Answer: Ephesians 4:32

ADDITIONS Answers
Old and New Testament

1. Book: 17 + 6 = 23
    Chapter: 5 + 9 = 14
    Verse: 3 + 0 = 3
    Answer: Isaiah 14:3

2. Book: 18 + 1 = 19
    Chapter: 28 + 43 = 71
    Verse: 7 + 1 = 8
    Answer: Psalm 71:8

3. Book: 21 + 19 = 40
    Chapter: 11 + 7 = 18
    Verse: 1 + 2 = 3
    Answer: Matthew 18:3

4. Book: 4 + 0 = 4
    Chapter: 7 + 5 = 12
    Verse: 3 + 3 = 6
    Answer: Numbers 12:6

5. Book: 14 + 10 = 24
    Chapter: 7 + 2 = 9
    Verse: 12 + 11 = 23
    Answer: Jeremiah 9:23

6. Book: 25 + 16 = 41
    Chapter: 5 + 9 = 14
    Verse: 50 + 22 = 72
    Answer: Mark 14:72

7. Book: 16 + 16 = 32
    Chapter: 0 + 3 = 3
    Verse: 4 + 1 = 5
    Answer: Jonah 3:5

8. Book: 3 + 3 = 6
    Chapter: 3 + 2 = 5
    Verse: 8 + 7 = 15
    Answer: Joshua 5:15

9. Book: 28 + 18 = 46
    Chapter: 8 + 4 = 12
    Verse: 3 + 9 = 12
    Answer: 1 Corinthians 12:12

10. Book: 35 + 7 = 42
    Chapter: 7 + 1 = 8
    Verse: 0 + 1 = 1
    Answer: Luke 8:1

11. Book: 15 + 6 = 21
    Chapter: 3 + 4 = 7
    Verse: 6 + 3 = 9
    Answer: Ecclesiastes 7:9

12. Book: 40 + 4 = 44
    Chapter: 3 + 0 = 3
    Verse: 2 + 4 = 6
    Answer: Acts 3:6

13. Book: 11 + 3 = 14
    Chapter: 16 + 13 = 29
    Verse: 7 + 8 = 15
    Answer: 2 Chronicles 29:15

14. Book: 3 + 0 = 3
    Chapter: 14 + 9 = 23
    Verse: 18 + 4 = 22
    Answer: Leviticus 23:22

15. Book: 20 + 25 = 45
    Chapter: 6 + 6 = 12
    Verse: 9 + 9 = 18
    Answer: Romans 12:18

16. Book: $8 + 4 = 12$
    Chapter: $3 + 5 = 8$
    Verse: $0 + 1 = 1$
    Answer: 2 Kings 8:1

17. Book: $6 + 4 = 10$
    Chapter: $2 + 7 = 9$
    Verse: $3 + 4 = 7$
    Answer: 2 Samuel 9:7

18. Book: $17 + 3 = 20$
    Chapter: $9 + 5 = 14$
    Verse: $0 + 1 = 1$
    Answer: Proverbs 14:1

19. Book: $16 + 27 = 43$
    Chapter: $3 + 2 = 5$
    Verse: $7 + 1 = 8$
    Answer: John 5:8

20. Book: $0 + 1 = 1$
    Chapter: $3 + 1 = 4$
    Verse: $9 + 2 = 11$
    Answer: Genesis 4:11

SUBTRACTIONS Answers
Old and New Testament

1. Book: 66 – 26 = 40
   Chapter: 27 – 7 = 20
   Verse: 40 – 12 = 28
   Answer: Mathew 20:28

2. Book: 50 – 27 = 23
   Chapter: 66 – 25 = 41
   Verse: 30 – 13 = 17
   Answer: Isaiah 41:17

3. Book: 85 – 41 = 44
   Chapter: 27 – 21 = 6
   Verse: 12 – 9 = 3
   Answer: Acts 6:3

4. Book: 50 – 31 = 19
   Chapter: 150 – 64 = 86
   Verse: 17 – 6 = 11
   Answer: Psalm 86:11

5. Book: 66 – 42 = 24
   Chapter: 25 – 12 = 13
   Verse: 30 – 7 = 23
   Answer: Jeremiah 13:23

6. Book: 60 – 14 = 46
   Chapter: 22 – 9 = 13
   Verse: 6 – 5 = 1
   Answer: 1 Corinthians 13:1

7. Book: 25 – 22 = 3
   Chapter: 40 – 14 = 26
   Verse: 20 – 7 = 13
   Answer: Leviticus 26:13

8. Book: 64 – 32 = 32
   Chapter: 8 – 4 = 4
   Verse: 17 – 15 = 2
   Answer: Jonah 4:2

9. Book: 60 – 12 = 48
   Chapter: 10 – 8 = 2
   Verse: 27 – 7 = 20
   Answer: Galatians 2:20

10. Book: 7 – 1 = 6
    Chapter: 18 – 12 = 6
    Verse: 42 – 22 = 20
    Answer: Joshua 6:20

11. Book: 39 – 38 = 1
    Chapter: 20 – 3 = 17
    Verse: 2 – 1 = 1
    Answer: Genesis 17:1

12. Book: 95 – 49 = 46
    Chapter: 45 – 30 = 15
    Verse: 66 – 14 = 52
    Answer: 1 Corinthians 15:52

13. Book: 39 – 18 = 21
    Chapter: 10 – 2 = 8
    Verse: 19 – 8 = 11
    Answer: Ecclesiastes 8:11

14. Book: 50 – 5 = 45
    Chapter: 21 – 8 = 13
    Verse: 13 – 3 = 10
    Answer: Romans 13:10

15. Book: 66 – 52 = 14
    Chapter: 15 – 14 = 1
    Verse: 22 – 10 = 12
    Answer: 2 Chronicles 1:12

16. Book: $25 - 13 = 12$
    Chapter: $40 - 28 = 12$
    Verse: $11 - 2 = 9$
    Answer: 2 Kings 12:9

17. Book: $66 - 23 = 43$
    Chapter: $14 - 8 = 6$
    Verse: $26 - 12 = 14$
    Answer: John 6:14

18. Book: $18 - 8 = 10$
    Chapter: $26 - 10 = 16$
    Verse: $2 - 1 = 1$
    Answer: 2 Samuel 16:1

19. Book: $35 - 8 = 27$
    Chapter: $9 - 3 = 6$
    Verse: $31 - 9 = 22$
    Answer: Daniel 6:22

20. Book: $60 - 40 = 20$
    Chapter: $20 - 12 = 8$
    Verse: $15 - 4 = 11$
    Answer: Proverbs 8:11

MULTIPLICATIONS Answers
Old and New Testament

1. Book: $41 \times 1 = 41$
   Chapter: $5 \times 3 = 15$
   Verse: $23 \times 2 = 46$
   Answer: Mark 15:46

2. Book: $2 \times 3 = 6$
   Chapter: $2 \times 2 = 4$
   Verse: $23 \times 1 = 23$
   Answer: Joshua 4:23

3. Book: $7 \times 3 = 21$
   Chapter: $3 \times 3 = 9$
   Verse: $5 \times 1 = 5$
   Answer: Ecclesiastes 9:5

4. Book: $2 \times 21 = 42$
   Chapter: $2 \times 2 = 4$
   Verse: $4 \times 2 = 8$
   Answer: Luke 4:8

5. Book: $7 \times 2 = 14$
   Chapter: $3 \times 8 = 24$
   Verse: $4 \times 5 = 20$
   Answer: 2 Chronicles 24:20

6. Book: $2 \times 2 = 4$
   Chapter: $8 \times 4 = 32$
   Verse: $9 \times 3 = 27$
   Answer: Numbers 32:27

7. Book: $23 \times 1 = 23$
   Chapter: $3 \times 4 = 12$
   Verse: $1 \times 2 = 2$
   Answer: Isaiah 12:2

8. Book: $4 \times 10 = 40$
   Chapter: $5 \times 5 = 25$
   Verse: $2 \times 5 = 10$
   Answer: Matthew 25:10

9. Book: $2 \times 19 = 38$
   Chapter: $4 \times 1 = 4$
   Verse: $2 \times 3 = 6$
   Answer: Zechariah 4:6

10. Book: $6 \times 4 = 24$
    Chapter: $2 \times 1 = 2$
    Verse: $1 \times 7 = 7$
    Answer: Jeremiah 2:7

11. Book: $2 \times 16 = 32$
    Chapter: $3 \times 1 = 3$
    Verse: $1 \times 3 = 3$
    Answer: Jonah 3:3

12. Book: $2 \times 22 = 44$
    Chapter: $4 \times 1 = 4$
    Verse: $6 \times 2 = 12$
    Answer: Acts 4:12

13. Book: $6 \times 2 = 12$
    Chapter: $1 \times 19 = 19$
    Verse: $5 \times 3 = 15$
    Answer: 2 Kings 19:15

14. Book: $19 \times 1 = 19$
    Chapter: $8 \times 15 = 120$
    Verse: $1 \times 1 = 1$
    Answer: Psalm 120:1

15. Book: $13 \times 5 = 65$
    Chapter: $1 \times 1 = 1$
    Verse: $7 \times 3 = 21$
    Answer: Jude 1:21

16. Book: $2 \times 23 = 46$
    Chapter: $4 \times 2 = 8$
    Verse: $3 \times 2 = 6$
    Answer: 1 Corinthians 8:6

17. Book: $5 \times 2 = 10$
    Chapter: $3 \times 7 = 21$
    Verse: $7 \times 1 = 7$
    Answer: 2 Samuel 21:7

18. Book: $8 \times 8 = 64$
    Chapter: $1 \times 1 = 1$
    Verse: $2 \times 1 = 2$
    Answer: 3 John 1:2

19. Book: $43 \times 1 = 43$
    Chapter: $1 \times 11 = 11$
    Verse: $22 \times 2 = 44$
    Answer: John 11:44

20. Book: $1 \times 1 = 1$
    Chapter: $8 \times 3 = 24$
    Verse: $17 \times 3 = 51$
    Answer: Genesis 24:51

DIVISIONS Answers
Old and New Testament

1. Book: $230 \div 10 = 23$
   Chapter: $52 \div 2 = 26$
   Verse: $16 \div 4 = 4$
   Answer: Isaiah 26:4

2. Book: $82 \div 2 = 41$
   Chapter: $30 \div 3 = 10$
   Verse: $57 \div 3 = 19$
   Answer: Mark 10:19

3. Book: $48 \div 2 = 24$
   Chapter: $64 \div 4 = 16$
   Verse: $17 \div 1 = 17$
   Answer: Jeremiah 16:17

4. Book: $36 \div 6 = 6$
   Chapter: $40 \div 5 = 8$
   Verse: $36 \div 2 = 18$
   Answer: Joshua 8:18

5. Book: $80 \div 2 = 40$
   Chapter: $56 \div 2 = 28$
   Verse: $36 \div 6 = 6$
   Answer: Matthew 28:6

6. Book: $56 \div 4 = 14$
   Chapter: $21 \div 3 = 7$
   Verse: $28 \div 2 = 14$
   Answer: 2 Chronicles 7:14

7. Book: $64 \div 2 = 32$
   Chapter: $21 \div 7 = 3$
   Verse: $30 \div 3 = 10$
   Answer: Jonah 3:10

8. Book: $43 \div 1 = 43$
   Chapter: $60 \div 4 = 15$
   Verse: $70 \div 10 = 7$
   Answer: John 15:7

9. Book: $48 \div 8 = 6$
   Chapter: $72 \div 3 = 24$
   Verse: $62 \div 2 = 31$
   Answer: Joshua 24:31

10. Book: $1 \div 1 = 1$
    Chapter: $82 \div 2 = 41$
    Verse: $100 \div 4 = 25$
    Answer: Genesis 41:25

11. Book: $92 \div 2 = 46$
    Chapter: $270 \div 27 = 10$
    Verse: $52 \div 4 = 13$
    Answer: 1 Corinthians 10:13

12. Book: $76 \div 4 = 19$
    Chapter: $236 \div 2 = 118$
    Verse: $32 \div 4 = 8$
    Answer: Psalm 118:8

13. Book: $63 \div 3 = 21$
    Chapter: $20 \div 2 = 10$
    Verse: $33 \div 3 = 11$
    Answer: Ecclesiastes 10:11

14. Book: $144 \div 12 = 12$
    Chapter: $100 \div 5 = 20$
    Verse: $40 \div 8 = 5$
    Answer: 2 Kings 20:5

15. Book: $132 \div 3 = 44$
    Chapter: $64 \div 8 = 8$
    Verse: $85 \div 5 = 17$
    Answer: Acts 8:17

16. Book: $36 \div 9 = 4$
    Chapter: $108 \div 3 = 36$
    Verse: $54 \div 6 = 9$
    Answer: Numbers 36:9

17. Book: $108 \div 4 = 27$
    Chapter: $21 \div 3 = 7$
    Verse: $150 \div 6 = 25$
    Answer: Daniel 7:25

18. Book: $84 \div 2 = 42$
    Chapter: $45 \div 9 = 5$
    Verse: $40 \div 2 = 20$
    Answer: Luke 5:20

19. Book: $60 \div 3 = 20$
    Chapter: $62 \div 2 = 31$
    Verse: $30 \div 3 = 10$
    Answer: Proverbs 31:10

20. Book: $95 \div 5 = 19$
    Chapter: $438 \div 3 = 146$
    Verse: $40 \div 5 = 8$
    Answer: Psalm 146:8